Leprosy and colonialism

Manchester University Press

SOCIAL HISTORIES OF MEDICINE

Series editors: David Cantor and Keir Waddington

Social Histories of Medicine is concerned with all aspects of health, illness and medicine, from prehistory to the present, in every part of the world. The series covers the circumstances that promote health or illness, the ways in which people experience and explain such conditions and what, practically, they do about them. Practitioners of all approaches to health and healing come within its scope, as do their ideas, beliefs and practices, and the social, economic and cultural contexts in which they operate. Methodologically, the series welcomes relevant studies in social, economic, cultural and intellectual history, as well as approaches derived from other disciplines in the arts, sciences, social sciences and humanities. The series is a collaboration between Manchester University Press and the Society for the Social History of Medicine.

Previously published

The metamorphosis of autism: A history of child development in Britain *Bonnie Evans*

The politics of vaccination: A global history *Edited by Christine Holmberg, Stuart Blume and Paul Greenough*

Payment and philanthropy in British healthcare, 1918–48 *George Campbell Gosling*

Leprosy and colonialism

Suriname under Dutch rule, 1750–1950

Stephen Snelders

Manchester University Press

Copyright © Stephen Snelders 2017

The right of Stephen Snelders to be identified as the author of this work has been asserted by him in accordance with the Copyright, Designs and Patents Act 1988.

Published by Manchester University Press
Altrincham Street, Manchester M1 7JA

www.manchesteruniversitypress.co.uk

British Library Cataloguing-in-Publication Data
A catalogue record for this book is available from the British Library

Library of Congress Cataloging-in-Publication Data applied for

ISBN 978 1 5261 1299 6 hardback

First published 2017

The publisher has no responsibility for the persistence or accuracy of URLs for any external or third-party internet websites referred to in this book, and does not guarantee that any content on such websites is, or will remain, accurate or appropriate.

Typeset by Out of House Publishing
Printed by Lightning Source

Contents

List of figures	vi
List of tables	vii
Acknowledgements	viii
Introduction	1
Part I Leprosy in a slave society	**19**
1 The making of a colonial disease in the eighteenth century	21
2 A policy of 'Great Confinement', 1815–1863	43
3 Slaves and medicine: black perspectives	78
4 'Battleground in the jungle': the Batavia leprosy asylum in the age of slavery	93
Part II Leprosy in a modern colonial state	**117**
5 Transformations and discussion: Suriname and the Netherlands, 1863–1890	119
6 Towards a modern colonial state: reorganizing leprosy care, 1890–1900	142
7 Developing modern leprosy politics, 1900–1950	161
8 Colonial medicine and folk beliefs in the modern era	199
9 Complex microcosms: asylums and treatments, 1900–1950	219
Conclusion	247
Sources and select bibliography	251
Index	270

Figures

1 Investigations of the Committee of Investigation,
  1831–1859 (Source: Landré, 'Naschrift', p. 233;
  Drognat Landré, Besmettelijkheid der lepra, p. 34)   60
2 Segregated sufferers in Batavia, 1849–1897 (Source:
  Colonial Reports)   104
3 Numbers of patients in the asylums, 1896–1949
  (Source: Colonial Reports; Public Health Service
  reports: GS 1258, 1262, 1264, BP T; Hallewas,
  'Gezondheidszorg', App. XI (voor na 1918); Majella
  1896–1898: BP T178)   166
4 Admitted and registered leprosy sufferers,
  1928–1949 (Source: Colonial Reports; GS 1262;
  Report Leprabestrijdingsdienst 1946, GS 1264)   182

Tables

1	Investigations of the Committee of Investigation, 1831–1859	59
2	Segregated sufferers in Batavia, 1849–1896	104
3	Numbers of patients in the asylums, 1896–1949	164
4	Admitted and registered leprosy sufferers, 1928–1949	183
5	Percentages of admitted and registered leprosy sufferers among the population, calculated according to the 1921 census	184
6	Percentages of admitted and registered leprosy sufferers among the population, recalculated according to the 1950 census	184

Acknowledgements

The research and writing of *Leprosy and Colonialism* was undertaken as part of the project 'Leprosy and Empire', a comparative history project investigating the management of leprosy in Dutch colonies in the East and West Indies (present-day Indonesia and Suriname) that was financed by NWO-Geesteswetenschappen (the Dutch Organization for Scientific Research, division of Humanities; file no. 360–69–020). This project would never have been conceived and financed without the initiative and continued impetus of Henk Menke. Henk Menke teamed up with Toine Pieters and ensured the support of the Descartes Center for the History and Philosophy of the Sciences and the Humanities at Utrecht University and built the foundation for the project proposal. Frank Huisman at the Julius Center for Health Sciences and Primary Care of the University Medical Center in Utrecht successfully submitted the final research proposal for financing to the NWO, supervised the project and was essential in bringing it to a successful close. The Surinamese research was conducted at the Julius Center, while the Indonesia research was performed by Leo van Bergen at the Royal Netherlands Institute of Southeast Asian and Caribbean Studies (KITLV) in Leiden, and supervised by Peter Boomgaard and Henk Schulte Nordholt. The progress and findings of the research were discussed in an advisory board where historians and dermatologists tried to bridge the gaps between the various disciplines and disparate styles of viewing medical and colonial history. Members of the board included Leo van Bergen, Peter Boomgaard, William Faber, Liesbeth Hesselink, Frank Huisman, Henk Menke, Toine Pieters, Henry de Vries, and the present author. The author wishes to acknowledge and thank all those

Acknowledgements

concerned for their participation and input. The results as presented in this book are, of course, solely his responsibility.

The author further wishes to thank the following people. Jane Buckingham and Jo Robertson were helpful in discussing the various themes and problems of the project and sharing insights and research findings. Michael Worboys shared his unpublished writings on the history of leprosy. In Suriname, historian Mildred Caprino told of the experiences of her family in Suriname with leprosy sufferers, and lent assistance and advice. Historians Maurits Hassankhan, Jerry Eggers, Henri Brug and Tanya Sitaram of the National Archive in Paramaribo, Chequita Goedschalk of the archive of the Bisdom Paramaribo, and J. van Putten of the Stichting Surinaams Museum in Paramaribo were also of great help. Melinda Reyme and Jack Menke of the Anton de Kom University in Paramaribo made their transcripts of interviews of former Hansen patients in Suriname available to the author. In the Netherlands, Tinde van Andel shared her expertise on Surinamese medical plants, and Joop Vernooij offered information and insights on Catholic priests and nuns in Suriname. Natalie Zemon Davis shared her discovery of Schilling's birthplace with the author and kindly discussed her research on Suriname. Kirsten Beukenkamp, who was working on a comparative history of leprosy and AIDS in the French and Dutch West Indies, Alice Cruz, Charlie van Genuchten, and Debbie McCollin all shared their findings with the author. The board members of Stichting Historia Medicinae kindly financed the text correction of the manuscript. Finally, many thanks to Keir Waddington of the Society for the Social History of Medicine for his insightful comments on the manuscript, and to Julia Challinor for the final editing.

Introduction

In the 2012 American animated comedy film *The Pirates! Band of Misfits*, the pirates attack and board a ship. To their horror, they are confronted with leprosy sufferers. One of the sufferers pulls off his arm and the pirates, aghast, beat a hasty retreat. Of course, this scene was not meant as a serious depiction of leprosy or Hansen's disease, as it is called today. However, patients who had formerly had Hansen's disease complained and the filmmakers hastily changed the leprosy ship into a plague ship. *The Pirates* film highlighted that leprosy's horrendous image remains still vibrant in Western culture, and the controversial nature of this image.

Those who suffer from leprosy have been historically stigmatized and excluded from society.[1] In attempts to understand these stigmatizing processes, the 'leprous body' has been conceptualized as the ultimate signifier of blurred boundaries between life and death. The British historian Rod Edmond draws on the work of anthropologist Mary Douglas and linguist/philosopher Julia Kristeva to theorize this 'leprous body'. For Edmond, leprosy in biblical times (not necessarily the same disease as modern leprosy) was an unclean abomination undermining the wholeness and completeness of the human body. Rituals and taboos were and are in place to protect the body's wholeness and to make a clear distinction and boundary between clean and unclean, order and disorder. However, in reality these distinctions are not so clear cut. To Edmond, the 'leprous body' is the most horrendous manifestation of the challenge of making clear distinctions: 'a mordant instance … death infecting life … something rejected from which one does not part'.[2]

Explaining how leprosy was considered in various historical settings by referring to categories of uncleanliness in antiquity, however, is problematic. Rather than taking a cue from a philosophical position on the wholeness of human nature and leprosy's abhorrent threat to this wholeness, in *Leprosy and Colonialism* I historicize how leprosy has been framed and addressed. Here leprosy is considered as a phenomenon shaped by time and place, and in particular by its relationship with colonialism.

Since the end of the nineteenth century, leprosy has been understood as a chronic infectious disease. Symptoms can take from ten to twenty years to develop and include anaesthesia (inability to feel pain) and inflammation of the skin, nerves, and eyes. Body parts do not fall off, but rather a weakening of the body's defences against secondary infections can lead to deformations and diseases of the extremities (fingers and toes). When repeated injuries occur, the inability to feel pain can lead to loss of extremity parts. Effective medication for leprosy only became available after the Second World War.

Although leprosy had ceased to be endemic across most of Europe by the early modern period, in the mid-eighteenth century Europeans encountered a disease they identified as leprosy in a completely new setting in another part of the globe among people of colour in Caribbean plantation colonies. From approximately 1750 onwards, leprosy or 'boasie' was seen by the Dutch rulers and Dutch colonial medicinal professionals in Suriname (the Dutch part of Guiana on the northern coast of South America), as an important danger to the slave population's health, public hygiene, and colonial rule. It was even feared that the disease might cross bloundaries and return to the Netherlands, thus undermining the global Dutch colonial empire.

Suriname was a Dutch construct. It was a plantation society where the vast majority of the population consisted of imported slaves from Africa, who had to be controlled. In this respect, Suriname was quite typical of other Caribbean plantation colonies. The Caribbean colonies specialized in exporting commodities, sugar in particular, using a system of coercion whereby coloured slaves (and after the abolition of slavery, Asian indentured labourers) were used as an agricultural labour force.[3] As historian Doris Garraway writes, 'The Caribbean plantation system … was founded on what was … the most brutal experiment in social engineering and physical repression.'[4] The colonial framing

Introduction

of leprosy has to be investigated and understood within the context of the plantation economy and the attempts to control and 'colonize' the bodies of the labourers – the slaves.[5] Slave medicine (medical care for and medical care among the slaves) became a focal point of contestation and control.[6]

By 1790, compulsory segregation polices for leprosy sufferers were in place. These policies continued long after the abolition of slavery in Suriname in 1863, and after the end of direct Dutch colonial rule in 1950.[7] After the emancipation of the slaves, the social and cultural heritage of slavery continued to exercise an influence on the history of leprosy. The legacy of leprosy control and the slave society's fear of the disease later affected how leprosy was viewed and addressed in the modernizing colonial state. This legacy continued in spite of the profound changes in Surinamese society, such as the large-scale immigration of indentured labourers from British India and the Dutch East Indies and the transformation of the plantation economy into late colonial capitalism. *Leprosy and Colonialism* investigates the history of leprosy in Suriname within the context of Dutch colonial power, slavery and its legacy, and racial conflict.

Historiography: leprosy and imperialism

The history of leprosy's connection with Caribbean plantation colonialism has received little attention from historians compared to its connections with the growth of Western imperialism in the nineteenth century.[8] A central focus of investigation has been the development of the notion of leprosy as an 'imperial danger' at the end of the nineteenth century and leprosy's connections to imperialism and Social Darwinism.[9] Leprosy has been perceived as circulating throughout European empires through the migration of non-white people and the circulation of goods, thus endangering white people. In an influential study published in 1989, Zachary Gussow concluded as follows:

> By the nineteenth century [leprosy] had reappeared and by the end of the century had caused Western nations to panic. During the period of nineteenth-century imperialism, the disease was discovered to be hyperendemic in those parts of the world that Western nations were annexing and colonizing. The discovery of leprosy in the colonial world,

and the excitement in the 1860s generated by the announcement of an epidemic in Hawaii, revived Western concerns about a disease that otherwise remained but a memory.[10]

Gussow related this 'rediscovery' and renewed fears of leprosy to anxieties about Chinese immigration and an endangerment of 'Americanness' in the United States. For Gussow, leprosy was framed as a disease of racially 'inferior' people. According to Gussow, the association of this rediscovered leprosy with biblical and medieval leprosy led to the stigmatization of the leprosy sufferers, their isolation, and segregation policies.

Thus, Gussow made explicit links between the stigmatization of leprosy and racial fears spreading worldwide at the end of the nineteenth century owing to international migration movements. Questions of health and disease were conflated and confused with political rhetoric and racial tensions. Historians have adopted this idea. For example, Jo Robertson has argued that in the Australian territory of Queensland in the 1890s, leprosy was racialized. For Roberston,

> An extraordinary discursive formation came into play that was about the colony being 'corrupt' both politically and also in terms of the disease leprosy ... The workers saw the importation of indentured labour undermining their hard won rights and they opposed them on the basis that the Polynesian and Melanesian labourers were, with political support, introducing disease (leprosy) into the colony.[11]

Historians have further directed special attention to the role of missionary societies in managing leprosy since the religious revival of the 1860s.[12] Addressing leprosy has also been situated in the context of the construction of national identities in the era of imperialism.[13] It is remarkable that an important part of the modern history of leprosy has remained insufficiently explored, conceptually as well as empirically, namely, its history in the eighteenth- and nineteenth-century Caribbean.[14]

Leprosy and race

In the eighteenth- and nineteenth-century Caribbean colonies, the identity of the supposed carriers of leprosy took central place in the framing of the disease. Colonial rulers in the eighteenth-century

Caribbean thought that a key risk group of carriers were their African slaves. The first constructions of leprosy as a danger to white dominance transmitted by an 'inferior' race, and as a disease similar or identical to biblical and medieval leprosy, began in the Caribbean. Hence, race is of key importance to the history of leprosy.

According to the historiography of colonial medicine, racism was on the increase after 1800. Mark Harrison has connected this increase to the history of slavery. To defend themselves against attacks on the slave trade, European colonizers emphasized their supposed fundamental biological difference with the Africans.[15] The idea of a fundamental difference between races developed within a colonial context. Historian Alfred Crosby showed in his seminal work on *The Columbian Exchange* that from the very first, the discoverers of the New World wondered about their differences with the indigenous inhabitants. Some Europeans entertained the notion of 'multiple creations': God might have created fundamentally distinct worlds, the Old and the New. To the eighteenth-century French naturalist Buffon it was clear that Amerindians or Native Americans were in all respects inferior to Europeans. Furthermore, colonizers observed that since the Conquest, diseases that had been prevalent among the inhabitants of one part of the world had begun to plague the inhabitants of other parts.[16] Kenneth Kiple and Richard Sheridan have described the epidemiological transitions and the changing disease environment in the Caribbean in the eighteenth century in more detail and highlighted changes related to the forced migration of Africans to the New World. Yellow fever, filariasis, malaria, and yaws were some of the diseases that became rampant on Caribbean islands and threatened the success of European military operations.[17] For Sheridan, 'Faced with numerous diseases that were indigenous to Africa ... attention [of European doctors] was directed to the differences between Africans and Europeans with respect to resistance and susceptibility to various diseases.'[18] The changing disease environment and the close proximity to slaves of African descent prompted inquiries into the health and disease of the non-white population in the Caribbean much earlier than in Asia.[19]

By the later eighteenth century, what Londa Schiebinger has called the 'anatomy of difference' between races was widely debated among

European scientists and savants. Explanations for these 'differences' ranged between environmentalism and hereditarianism, including combinations of both.[20] While in Europe this was more of a theoretical concern, in the colonies the question of why and to what extent various races were prey to specific diseases was of eminent practical concern. As Sean Quinian writes in his study of the French colonies, colonial doctors had to find an explanation for the 'selective nature of disease' since they observed that Africans and Europeans, 'responded quite differently to the exigencies of the Caribbean tropics ... In contrast to physicians in Europe (who emphasized differences of class), colonial doctors frequently stressed biological differences of a racial type.'[21] According to Quinian, it was a French physician, Pierre Barrère, who was one of the first to identify a 'morbid otherness' among the African population in 1741. To Barrère (who had spent five years working in Cayenne, the neighbouring French colony to Suriname), Europeans considered the African as a source of pollution.[22] The ultimate distinction between the races was located in the amount of self-control a male European could exert to regulate his functioning in accordance with the environment. 'In a sense, the diseased body became the ultimate signifier of not just the pathological milieu but the total lack of physical self-control exercised by the European individual', writes Quinian.[23] Differences in 'passions of the mind' were used to explain racial differences in disease patterns.

In Suriname, leprosy became a focus of ideas of racial difference, the failure of making and upholding clear distinctions between racial boundaries, and a threat to the Europeans that could easily extend to Europe. These fears led to early local compulsory segregation policies rather than policies that spread 'outward' from a colonial or imperialist 'centre' to the periphery of empire.[24] The policies were developed from the perspective of a 'slaveholder's knowledge' as long as it is understood that 'slaveholders' were not only the actual slave owners, but also 'many more with a direct or indirect interest in slaveholding through family connections or professional and business arrangements'.[25] Hence, addressing leprosy in Suriname became integral to what historian David Arnold has called the 'colonization of the body' or the conflict over who had the right to control whose body.[26]

Leprosy politics in Suriname

Compulsory segregation policies began in Suriname in the second half of the eighteenth century and anticipated global developments in the age of imperialism. The policies took the form of a 'Great Confinement' (to borrow a phrase from Michel Foucault) in the decades between 1830 and 1860.[27] Close to one out of every 100 inhabitants were condemned or suspected of having leprosy, and confined to the Batavia leprosy asylum or segregated in their own homes or elsewhere. Although segregation policies seemed to be ebbing after the abolition of slavery in 1863, colonial leprosy control at the end of the nineteenth century gave segregation a new impetus. 'Modern Dutch' colonial policies in Suriname were characterized by the combination of authoritarianism with a belief in rational order, linear progress, and standardized conditions of knowledge. Colonial health policies became 'modern' in this sense, which affected leprosy control especially after the 1890s. Thus, segregation policies for leprosy can be understood as an attempt at social engineering and described as 'authoritarian modernist', which is a useful term for distinguishing the pre- and post-emancipation colonial state.[28]

In Suriname, the difference between the 'old' leprosy asylums founded in the age of slavery, Voorzorg and Batavia, on the one hand, and the modern leprosy asylums of the twentieth century, Groot-Chatillon, Majella, and Bethesda, on the other, is exemplary of modern colonial health policies. The first asylums were more or less dumping grounds of villages where whole families lived excluded from society with relatively reasonable freedom of movement, but little medical care. The modern asylums had relatively improved hygienic and medical conditions, but freedom of movement was limited and inmate discipline increased. This was part of what Dutch doctors claimed was a change from a coercive to a medical leprosy policy. If the reality was more complex, this shift away from a slave holder's perspective seems to fit with Suriname's transition to a more 'modern' colonial state. However, the shift in leprosy policies was not a total change: modern colonial society continued the heritage of framing leprosy that originated within the old colonial slave society.

Modern leprosy politics also continued the heritage of the role of missionary societies in the fight against leprosy. Historians have focused attention on Christian and especially Protestant missionaries in the fight against leprosy in the British Empire and elsewhere. Michael Worboys has written about the role played by Christian missionary healthcare (together with medical humanism and colonial developmental policies) in the construction and implementation of policies that aimed to improve the population's welfare while realizing an imperial 'mission'. Within a framework whereby Christianity was propagated alongside a Western scientific rationalism and 'mandate' (strengthening the empire), leprosy was framed as the archetypical tropical disease prevalent among the races of colour. Western expertise was needed to fight this disease, and Christian missionaries were essential to implement their Western expertise.[29] In Suriname, in the 1820s, almost three-quarters of a century earlier, Catholic missionaries had already been given a central role in the fight against leprosy. Thus, they had demonstrated their essential role in the care and control of the Afro-Surinamese population to both the colonial state and the Catholics in the Netherlands who financed their missions.[30] The activities of Dutch Catholic priests in the Surinamese leprosy asylums were ahead of those of Protestant missionaries from the British Empire, and the activities of their internationally more famous colleague, Father Damien in the Kulawao leprosy settlement on the Hawaiian island of Molokai.[31] Here, as in the introduction and execution of compulsory segregation policies, Suriname anticipated global developments in the later nineteenth century.

Reconstructing the agency of leprosy sufferers

The colonial framing of leprosy and the development of leprosy politics by colonial medicine took place in a context of power relationships of the colonial state and colonial medicine on one side and leprosy sufferers, their kin, and their social groups on the other side. Historians have begun to focus on the complexities in the outcomes of encounters between Western medicine and non-Western contexts.[32] Authors such as Eric Silla, Jane Buckingham, and Keri Ingliss have shown how to bring the experiences and agency of sufferers in Africa and the Pacific to the centre of leprosy asylum narratives.[33] In Dutch Suriname and other regions, historians have to address the silences in colonial sources about the sufferers' experiences and agency. Everything that can be read in period sources or reliably traced back to these sources offers perspectives that are filtered through the eyes of European observers. For instance, Afro-Surinamese perspectives on disease and healing are distorted in this way, which is a typical example of the role of colonial power in the production and writing of history.[34] As Peter Hulme suggests, 'The only evidence that remains … are the very European texts that constitute the discourse of colonialism.'[35] In researching and writing *Leprosy and Colonialism*, strategies have been sought to break through these silences and distortions to avoid a limited and Eurocentric view in line with those historians who have shown that an alternative perspective can be taken with promising results by using extant colonial sources and reading them from a more 'bottom-up' perspective.[36]

This bottom-up perspective is of crucial importance in the investigation of the various aspects of leprosy politics in Dutch Suriname, such as the functioning of compulsory segregation and the population and patients' compliance, asylum functioning, and the problem of stigmatization. Rosemarijn Hoefte has described twentieth-century Suriname as a 'culture of domination and contestation'.[37] This applies to the eighteenth and nineteenth centuries as well. Contestation can take the form of resistance on the level of what anthropologist James Scott has called 'infrapolitics' or the 'hidden transcripts' of resentment and discontent (hidden, because of the lack of articulated media attention at the time) lying beneath or below ('infra') the articulated political sphere.[38] One example of contestation is the Afro-Surinamese and other ethnic groups' cultural resistance against the acceptance of Dutch religious and medico-scientific beliefs, and the continued survival and the importance of their folk beliefs. So too are the Afro-Surinamese refusals to cooperate with segregation politics and leprosy asylum patients' non-compliance. Reading the colonial sources from new perspectives, top-down as well as bottom-up, allows for the reconstruction of these dynamics of power, domination, and contestation.

In the 1990s and 2000s, historians such as Ruth Smith-Kipp, Warwick Anderson, and Rod Edmond analysed the asylums from the perspective of top-down control, and were influenced by Ervin Goffman's notion of a 'total institution' wherein the patient's behaviour and outlook are refashioned, and by the ideas of Michel Foucault.

Whereas Foucault had seen medieval leprosy colonies as an example of sovereign power, exile, and the enclosure of an abandoned marginalized group, these historians suggested that the modern leprosy asylums could be seen as an example of disciplinary power in which modern notions of citizenship were applied and patients were held under constant surveillance.[39] More recently, Jo Robertson, Jane Buckingham, and Kerri Ingliss have advocated a more nuanced approach, showing the variations, complexities, and contingencies in leprosy asylums, and reconstructing asylums as places where people could build a new sense of identity and community.[40] While the one perspective might be as 'true' as the other, reading the sources from the patient's perspective in the asylums is essential.

A bottom-up perspective is also needed for investigating stigmatization. By the end of direct Dutch colonial rule in Suriname, the problem of stigma had become a major cause of concern for medical practitioners treating leprosy. In 1951, Eugene R. Kellersberger, a leprosy doctor in the Belgian Congo and organizer of the first supplies of sulfone drugs in Suriname, claimed that there could be no medical hope for the patient with leprosy until the stigma of the disease was first removed.[41] Dutch anthropologist Annemarieke Blom conducted a series of interviews in 2001 and 2002 in Suriname with sixteen people who had had Hansen's disease and who were between twenty and eighty-seven years of age. She concluded that every one of them felt stigmatized for at least one or more reasons. Stigmatization was often connected with religious ideas; for instance that the Devil had cursed one's family or one had transgressed a taboo. Stigmatization was also a consequence of fears of infection by others in the environment, visible physical mutilations from the disease, and the connection between leprosy and poverty (or low social status).[42] However, we cannot assume a priori that these conclusions are valid for the entire history of leprosy in Suriname. In Thailand, Liora Navon concluded,

> prior to the discovery of a cure for [leprosy] its sufferers encountered ambivalent rather than severely stigmatizing reactions. Yet the public's selective exposure – mainly to beggars with the disease – paved the way to the perception of leprosy as the epitome of stigmatization and to its transformation into a metaphor for degradation.[43]

Similarly, L. K. Seng claims that many Chinese families and the larger public in British Singapore and Malaysia were quite sympathetic to leprosy sufferers before the start of compulsory segregation in 1897. It was this policy of compulsory segregation, as well as the general acceptance of the theory of the contagiousness of leprosy, that supposedly 'forged a new social horror' towards the disease.[44] In colonial Suriname, an iatrogenic stigma was framed by colonial medicine. However, it was particularly related to the sufferer's social status and/or ethnic background, rather than solely to the disease. An Afro-Surinamese leprosy sufferer could be looked down upon, but there is not much evidence that this was also the case with European leprosy sufferers unless there had been sexual relations with African women. Reading sources from a bottom-up perspective is essential for making sense of and historicizing the development of stigmatization.

Contents

This study traces the history of leprosy in Suriname in the context of the transformation of slave society and the modern colonial state, while reading historical sources from both the perspectives of the colonial rulers (top-down) and the ruled (bottom-up).[45] Part I considers leprosy in a slave society. Chapter 1 investigates the history of leprosy in eighteenth-century Suriname and the early colonial framing of the disease in the context of the slave economy. Chapter 2 presents the development and implementation of an intensified regime of detection and compulsory segregation after the Napoleonic wars that resulted in the leprosy edict of 1830 and the period of 'Great Confinement' of those with leprosy. Chapters 1 and 2 ask how and why these early examples of compulsory segregation policies came to be, how thorough and effective they were, and what their relationship was with the institute of slavery. The chapters also describe 'white' medical perspectives and practices as contrasted with a bottom-up perspective of 'black' beliefs and practices examined in Chapter 3. After 1824, patients with leprosy were sent to the Batavia leprosy asylum where only limited medical care was available, but more extensive materiel and spiritual care were provided by Catholics. Chapter 4 addresses the micro-cosmos of and power relations in Batavia.

Part II deals with the modern colonial state after the abolition of slavery in Suriname in 1863. After emancipation, interest in the problem of leprosy diminished for a time in Suriname, although there were fears in the Netherlands about a return of leprosy from the colonies. Chapter 5 investigates a period of transformation and discussion until 1890 about leprosy related to the end of the slave economy. Compulsory segregation received new impetus in the 1890s. Chapter 6 presents a reorganization of leprosy care in a modernizing colonial state. Modernization included both an emphasis on medical treatment and humanitarian care in new leprosy asylums and a new political accommodation in which the Protestants joined the Catholics in leprosy care. Chapter 7 investigates the changes in leprosy politics related to changes in modern colonial Suriname in the first half of the twentieth century. The leprosy edict of 1929 inaugurated a modernized and 'medicalized' leprosy politics that made outpatient treatment possible, but increased the detection and segregation of sufferers. One of the major problems for colonial medicine was the continued non-cooperation and non-compliance of sufferers who held on to their own beliefs and practices. As Chapter 8 shows, modern colonial medicine was interested in these local beliefs and tried to find ways to address them to ensure increased cooperation, especially from the Afro-Surinamese. Chapter 9 investigates the care and treatment of leprosy in the modern era and questions how and to what extent disciplining sufferers in modern asylums took place and succeeded. The conclusion then returns to the problem of leprosy within colonial power relations.

By investigating leprosy in Suriname, this book seeks to understand the complex reciprocities between knowledge, attitudes and practices towards leprosy over time, the agency of those with leprosy, and the ways in which colonial health policies came into being. In doing so, this book investigates the Caribbean origins of modern framing and management of leprosy; these origins have so far been neglected in the historiography of colonial and imperial medicine.

Notes

1 In this volume the word 'leprosy sufferers' is used throughout and not '(ex-) Hansen (disease) patients' or 'People Affected by Leprosy' (PAL). The term 'Hansen's disease' only came in vogue after the historical period this book is dealing with and its use would be anachronistic. It would also imply an answer to a question that is to a large extent unanswerable; namely whether

the people who were diagnosed with leprosy in the period under study really suffered from this disease as it is now understood. The word 'leprosy sufferer' is also anachronistic, but preferable to the English word 'leper' that has distinct negative connotations. 'People (or Individuals) Affected by Leprosy' is rather a mouthful, while using acronyms as 'PAL' and 'PALs' reads rather strange in a historical study. A general and popular introduction to the modern history of leprosy is T. Gould, *Don't Fence Me In: Leprosy in Modern Times* (London: Bloomsbury, 2005).

2   R. Edmond, *Leprosy and Empire: A Medical and Cultural History* (Cambridge: Cambridge University Press, 2006), p. 3.

3   On the history of Suriname, see R. van Lier, *Frontier Society: A Social Analysis of the History of Surinam* (The Hague: Martinus Nijhoff, 1971); H. Buddingh', *De geschiedenis van Suriname* (Amsterdam: Nieuw Amsterdam, 2012). General overviews of Caribbean plantation colonies: G. Heuman, *The Caribbean* (London: Hodder Arnold, 2006); J. Rogozinski, *A Brief History of the Caribbean: From the Arawak and the Carib to the Present* (New York: Facts on File, 1999). On plantation colonialism in Suriname: M. Schalkwijk, 'The plantation economy and the capitalist mode of production', in M. Schalkwijk and S. Small (eds.), *New Perspectives on Slavery and Colonialism in the Caribbean* (The Hague: Hamrit/Ninsee, 2012), pp. 14–40.

4   D. Garraway, *The Libertine Colony: Creolization in the French Caribbean* (Durham, NC: Duke University Press, 2005), p. 6.

5   D. Arnold, *Colonizing the Body: State Medicine and Epidemic Disease in Nineteenth-Century India* (Delhi: Oxford University Press, 1993).

6   On slave medicine, see T. L. Savitt, *Medicine and Slavery: The Diseases and Health Care of Blacks in Antebellum Virginia* (Urbana, IL: University of Illinois Press, 1978); S. M. Fett, *Working Cures: Healing, Health, and Power on Southern Slave Plantations* (Chapel Hill, NC: The University of North Carolina Press, 2002).

7   Suriname received formally autonomy in internal affairs in 1954 but direct Dutch rule from the Netherlands ceased in 1950. See J. C. Brons, *Het rijksdeel Suriname* (Haarlem: Bohn, 1952). In 1975 the country became an independent republic.

8   Z. Gussow, *Leprosy, Racism, and Public Health: Social Policy in Chronic Disease Control* (Boulder, CO: Westview Press, 1989); M. Vaughan, *Curing Their Ills: Colonial Power and African Illness* (Cambridge: Polity Press, 1991), pp. 77–99; J. Buckingham, *Leprosy in Colonial South India: Medicine*

and *Confinement* (Basingstoke: Palgrave, 2002); J. Robertson, 'Leprosy and the elusive *M. Leprae*: Colonial and imperial medical exchanges in the nineteenth century', *Manguinhos*, 10; suppl.1 (2003), pp. 13–40; Gould, *Don't Fence Me In*; W. Anderson, *Colonial Pathologies: American Tropical Medicine, Race and Hygiene in the Philippines* (Durham: Duke University Press, 2006), pp. 161–75; Edmond, *Leprosy and Empire*; J. Robertson, 'The leprosy asylum in India, 1886–1947', *Journal for the History of Medicine and Allied Sciences* 64 (2009), pp. 474–517; and other studies mentioned in A. Bashford, *Imperial Hygiene: A Critical History of Colonialism, Nationalism and Public Health* (Basingstoke: Palgrave Macmillan, 2004), pp. 81–93.
9 *Leprosy, an Imperial Danger*, is the title of a 1889 tract by Henry Press Wright (London: Churchill, 1889).
10 Gussow, *Leprosy, Racism, and Public Health*, p. 19.
11 Jo Robertson, email communication to the author, 18 February 2014. See J. Robertson, 'In a State of Corruption: Loathsome Disease and the Body Politic' (PhD thesis, University of Queensland, 1999), http://espace.library.uq.edu.au/view/UQ:193252/the13742.pdf [accessed on 21 October 2014]; Bashford, *Imperial Hygiene*; W. Anderson, *The Cultivation of Whiteness: Science, Health, and Racial Destiny in Australia* (Durham, NC: Duke University Press, 2006).
12 M. Worboys, 'The colonial world as mission and mandate: Leprosy and empire, 1900–1940', *Osiris* 15 (2000), pp. 207–18.
13 D. Obregon, 'Building national medicine: Leprosy and power in Colombia, 1870–1910', *Social History of Medicine* 15 (2002), pp. 89–108; A. Ki Che Leung, *Leprosy in China: A History* (London: Columbia University Press, 2008).
14 A modern history of leprosy in the Caribbean does at present not exist. On leprosy in Trinidad and Tobago: D. McCollin, 'Chacachacare: The island of lepers, 1922–1979', in C. Bonfield, J. Reinarz and T. Huguet-Termes, *Hospitals and Communities, 1100–1960* (Oxford: Peter Lang, 2013), pp. 263–90.
15 M. Harrison, 'The tender frame of man: Disease. climate, and racial differences in India and the West Indies', *Bulletin of the History of Medicine* 70 (1996), pp. 68–93; M. Harrison, *Medicine in an Age of Commerce and Empire: Britain and Its Tropical Colonies 1660–1830* (Oxford: Oxford University Press, 2010), p. 287.
16 A. W. Crosby, Jr., *The Columbian Exchange: Biological and Cultural Consequences of 1492* (Westport, CT: Greenwood, 1972), pp. 4–31.

17 K. F. Kiple, *The Caribbean Slave: A Biological History* (Cambridge: Cambridge University Press, 1984); R. B. Sheridan, *Doctors and Slaves: A Medical and Demographic History of Slavery in the British West Indies, 1680–1834* (Cambridge: Cambridge University Press, 1985). See also K. F. Kiple and Kriemhild Coneé Ornelas, 'Race, war and tropical medicine in the eighteenth-century Caribbean', in D. Arnold (ed.), *Warm Climates and Western Medicine: The Emergence of Tropical Medicine, 1500–1900* (Amsterdam: Rodopi, 1996), pp. 65–79; J. Handler, 'Diseases and medical disabilities of enslaved Barbadians, from the seventeenth century to around 1838', *The Journal of Caribbean History*, 2006, 1–38, pp. 177–214; J. R. McNeill, *Mosquito Empires: Ecology and War in the Greater Caribbean, 1629–1914* (Cambridge: Cambridge University Press, 2010).
18 Sheridan, *Doctors and Slaves*, p. 18.
19 M. Worboys, 'Tropical diseases', in W. F. Bynum and R. Porter (eds.), *Companion Encyclopedia of the History of Medicine*, vol. 2 (London: Routledge, 1993), pp. 512–36, on p. 517, on India: 'In fact, colonialism had little or no knowledge of health and disease amongst indigenous peoples.'
20 L. Schiebinger, 'The anatomy of difference: Race and sex in eighteenth-century science', *Eighteenth-Century Studies* 23 (1989/1990), pp. 387–405.
21 S. Quinian, 'Colonial encounters: Colonial bodies, hygiene and abolitionist policies in eighteenth-century France', *History Workshop Journal* 42 (1996), pp. 107–26, on p. 107.
22 Quinian, 'Colonial encounters', pp. 112–13.
23 Quinian, 'Colonial encounters', p. 120. According to male European savants, female Europeans also lacked the self-control of the male, as did Africans. (Schiebinger, 'Anatomy of difference'.)
24 Rather than introducing a concept of leprosy from Europe to the colonies, as for instance in Cambodia: S. Au, *Mixed Medicines: Health and Culture in French Colonial Cambodia* (Chicago, IL: The University of Chicago Press, 2011), pp. 157–79. On the analytical conceptualization of 'centre' and 'periphery': R. MacLeod (ed.), *Nature and Empire: Science and the Colonial Enterprise*, Osiris 15 (2000), pp. 1–317. For their relevance to the history of leprosy in Suriname: H. Menke, S. Snelders, and T. Pieters, 'Omgang met lepra in 'de West' in de negentiende eeuw. Tegendraadse maar betekenisvolle geluiden vanuit Suriname', *Studium* 2 (2009), pp. 65–77, on pp. 66–7.
25 E. Fox-Genovese and E. D. Genovese, *The Mind of the Master Class: History and Faith in the Southern Slaveholders' Worldview* (Cambridge: Cambridge University Press, 2005), p. 1.

26 Arnold, *Colonizing the Body*; see also N. T. Jensen, *For the Health of the Enslaved: Slaves, Medicine and Power in the Danish West Indies, 1803–1848* (Copenhagen: Museum Tusculaneum Press, 2012).
27 M. Foucault, *Madness and Civilization: A History of Insanity in the Age of Reason* (London: Routledge, 2001).
28 J. C. Scott, *Seeing Like A State: How Certain Schemes to Improve the Human Condition Have Failed* (New Haven, CT: Yale University Press, 1998), pp. 88–97, 177.
29 Worboys, 'Colonial world as mission and mandate'.
30 The word 'Afro-Surinamese' is used throughout the book for the whole period of study to avoid confusion and refers to the descendants of African slaves. This use includes the Maroons, the descendants of runaway slaves in the interior, and what today are called the Creoles. The word 'Creoles' refers now to descendants of the slaves apart from the Maroons, but it was used in the eighteenth and nineteenth centuries to refer to descendants of Europeans born in Suriname.
31 Edmond, *Leprosy and Empire*, p. 92.
32 R. Peckham and D. M. Pomfret (eds.), *Imperial Contagions: Medicine, Hygiene, and Cultures of Planning in Asia* (Hong Kong: Hong Kong University Press, 2013).
33 E. Silla, *People Are Not The Same: Leprosy and Identity in Twentieth Century Mali* (Portsmouth: Heinemann, 1998); Buckingham, *Leprosy in Colonial South India*; J. Buckingham, 'The inclusivity of exclusion: Isolation and community among leprosy-affected people in the South Pacific', *Health and History* 13 (2011), pp. 65–83; K. A. Ingliss, *Disease and Displacement in Nineteenth-Century Hawai'i* (Honolulu, HI: University of Hawai'i Press, 2013).
34 M.-R. Trouillot, *Silencing the Past: Power and the Production of History* (Boston, MA: Beacon Press, 1995).
35 P. Hulme, *Colonial Encounters: Europe and the Native Caribbean, 1492–1797* (London: Methuen, 1986), p. 8. Compare also Garraway, *Libertine Colony*, pp. 10–17.
36 K. K. Weaver, *Medical Revolutionaries: The Enslaved Healers of Eighteenth-Century Saint Domingue* (Urbana, IL: University of Illinois Press, 2006); P. F. Gómez Zuluaga, 'Bodies of encounter: Health, illness and death in the early modern African-Spanish Caribbean' (PhD thesis, Vanderbilt University, 2010); P. F. Gómez, 'The circulation of bodily knowledge in the seventeenth-century black Spanish Caribbean', *Social History of Medicine* 26 (2013), pp. 383–402.

Introduction   17

37 R. Hoefte, *Suriname in the Long Twentieth Century: Domination, Contestation, Globalization* (New York: Palgrave MacMillan, 2014), p. 2.
38 J. C. Scott, *Domination and the Art of Resistance: Hidden Transcripts* (New Haven: Yale University Press, 1990), p. 198. On open and hidden resistance in the Caribbean, see M. S. Hassankhan, B. V. Lal, and D. Munro (eds.), *Resistance and Indian Indentured Experience: Comparative Perspectives* (New Delhi: Manohar, 2014).
39 R. Smith Kipp, 'The evangelical uses of leprosy', *Social Science and Medicine* 39 (1994), pp. 165–78; W. Anderson, 'Leprosy and citizenship', *Positions* 6 (1998), pp. 707–30; Anderson, *Colonial Pathologies*, pp. 158–79; S.L. Burns, 'From 'leper villages' to leprosaria: Public health, nationalism and the culture of exclusion in Japan', in C. Strange and A. Bashford (eds.), *Isolation: Places and Practices of Exclusion* (London: Routledge, 2003), pp. 104–18; Bashford, *Imperial Hygiene*, pp. 81–113.
40 Robertson, 'Leprosy asylum', p. 474; Buckingham, 'Inclusivity of exclusion'; Ingliss, *Disease and Displacement*.
41 Cit. in Gussow, *Leprosy, Racism, and Public Health*, p. 4.
42 A. Blom, 'Angst voor lepra', *OSO* 22 (2003), pp. 90–8; A. Blom, 'Lepra in Suriname', *MensenBeelden* 6 (2004), pp. 28–31. On stigmatization in Suriname, see also P. Spapens, *Gwasi siki. Levensverhalen van Surinaamse mensen die lepra hebben gehad* (Tilburg: Pix4Profs, 2012); M. Reyme, 'Give ex-leprosy patients a voice' (unpublished manuscript, Anton de Kom Universiteit van Suriname, Paramaribo, 2013–2014).
43 L. Navon, 'Beggars, metaphors, and stigma: A missing link in the social history of leprosy', *Social History of Medicine* 11 (1998), pp. 89–106, on p. 102.
44 L. K. Seng, *Making and Unmaking the Asylum: Leprosy and Modernity in Singapore and Malaysia* (Petaling Jaya: Strategic Information and Research Development Centre, 2009), p. 30.
45 For earlier studies of leprosy in Suriname, taking different perspectives: E. van der Kuyp, 'De geschiedenis van lepra in Suriname tot 1971', *Surinaams Medisch Bulletin* 14 (1999), vol. 1, pp. 43–64; vol. 2, pp. 36–55; W. Hoogbergen and H. Ramsoedh (eds.), 'Lepra in Suriname', *OSO* 22 (2003), pp. 1–123; Menke, Snelders, and Pieters, 'Omgang met lepra'.

PART I
LEPROSY IN A SLAVE SOCIETY

1
The making of a colonial disease in the eighteenth century

On 19 August 1755, the Swedish botanist Daniel Rolander was exploring plants and wildlife near a coffee plantation on the banks of the Perica river in Suriname when he came across a horrendous sight. He wrote the following in his journal:

> [We] passed a few sylvan huts of blacks, where some sick unfortunates addressed us from afar, warning us not to approach any closer or enter their huts, because they said they were suffering from a contagious disease called 'boise' [*boasie* or *boassie*]. This disease is exanthematic and reminds one in some way of the leprosy of old and elephantism; it feeds upon the joints, and propagates via physical contact. Blacks infested with this disease are automatically relegated to a remote corner of the plantation, where they serve as guards and spend the rest of their time alienated from friends to keep the entire servile throng from contracting it.[1]

The city physician of the Surinamese capital Paramaribo told Rolander that African slaves had brought the disease over the ocean from Guinea in West Africa.[2] 'Boasie' was supposed to be the name of the place in Africa where the disease had come from.[3] By 1755, it was feared that boasie, quickly identified with leprosy, would spread from Africa via Suriname to Europe. African slaves were thought to carry the disease across the Atlantic to the Caribbean, where Europeans were then infected. Europeans could then in turn bring leprosy back to the Netherlands, where it had become extinct. To many observers, the health of the Dutch colonial and commercial empire was at stake.

This chapter argues that to the colonial rulers, boasie or leprosy's first manifestations in the eighteenth-century Caribbean, came to represent not only a threat to the health of the slave population and public hygiene, but also to the Netherlands and the Dutch colonial empire as a whole. Although, the definition, symptomatology, and aetiology of the disease still had to be developed in the eighteenth century, contemporaries routinely equated boasie with *elephantiasis graecorum* or *lepra arabum*, that is, leprosy (or in Dutch *melaatschheid*), the dreaded disease of the Middle Ages.[4] In the eighteenth century, many doctors and laymen regarded the disease as highly contagious.[5] Rolander noticed that in Suriname, healthy persons were advised to stay as far away from the sufferers as possible, not to enter their dwellings or touch them, and not to breathe the same air.[6] To many slave holders, leprosy seemed to endanger the health of their labour force and hence the functioning of the Surinamese slave society. Hence, boasie was not solely perceived as a medical problem, but came to be framed as a problem of geography and race, an economic problem disrupting the slave economy, and a socio-political threat. Knowledge was needed to counter this perceived threat, which resulted in a colonial framing of leprosy that influenced the perceptions and management of leprosy in the nineteenth and twentieth centuries, and leading to quite early policies of compulsory segregation at the end of the eighteenth century.

A Dutch slave doctor played a key role in the construction and implementation of slave holders' knowledge of leprosy. Godfried Wilhelm Schilling (1733–1734 – after 1795) had the idea that boasie was an African disease threatening the health of Europeans.[7] In 1769, he wrote that the 'Abyssinians' (by which he meant black Africans) had brought the disease to America. He saw almost no cases of boasie among the 'Aborigines' (the Native Americans), and so believed that the disease had not existed in the Americas before the immigration of African slaves. Therefore, Europeans who had physical contact with Africans were in danger of becoming contaminated as well. Since more and more slaves were coming to Suriname, it was to be expected that the incidence of the disease would increase.[8]

As one of the few physicians in the colony of Suriname, and since he conducted the medical examinations of newly arrived slaves, Schilling played an important part in framing health policy measures around boasie in the 1770s to 1790s. A closer investigation of his work demonstrates that the need for medical knowledge of the disease was driven by its effect on the slave trade and the slave economy in Suriname. The driving forces and profit motives of the slave economy profoundly shaped the aims, methods, and personnel involved in the search for medical knowledge in the tropics. Ultimately, profits were at stake if changes in the disease environment were not met by changes in medical practice and healthcare, for which medical knowledge was needed. Schilling delivered the medico-scientific 'evidence' and underpinnings of a public health policy of isolation and segregation that was implemented in Suriname in the second half of the eighteenth century. A key role in the development of this policy was his formulation of a racial pathology used in the understanding of leprosy's aetiology and epidemiology and in determining the measures that needed to be taken.

Suriname in the eighteenth century

The English had ceded Suriname to the Dutch in 1667. Apart from the Native Americans or Amerindians, just over 1,000 people lived in the colony, and among them 700 slaves. The colony came under the control of a private company, the Society of Suriname, composed of members of the Dutch West India Company (WIC), the city of Amsterdam, and a private investor. Under their control, the Dutch created a wealthy plantation economy based on African slave labour. By 1754, there were almost 1,500 Europeans (including many Jews who had settled there since the 1660s) and more than 33,000 slaves in Suriname. In 1783, there were probably more than 400 plantations, cultivating and exporting sugar, coffee, cacao, cotton, and timber. The plantations stretched along the rivers flowing from the Amazonian jungle into the Atlantic so from west to east: the Corantijn, Coppename, Saramacca, Suriname, Marowijne, and Amanibo rivers.

In 1783, the population had grown to over 2,000 Europeans and more than 50,000 slaves and included approximately 500 free coloured people and mulattos.[9] The Native Americans, mainly Carib and to a lesser degree Arawak Indians, lived in the jungle interior. They comprised a few thousand families who had signed peace treaties with the Dutch in 1686. These treaties safeguarded the Indians from slavery. More belligerent were the Maroons, fugitive African slaves from the plantations who had created their own tribes and societies in the interior: the Ndyuka, the Saramaccans, the Matawai, and the Boni. Their total number is estimated to have been between 6,000 and 7,000 from 1738 to 1786. They made their living mainly from forestry and developed their own Afro-Surinamese culture and religion. Eighteenth-century Suriname witnessed a fierce and protracted guerrilla war between Maroons and the army of the Society of Suriname. Peace treaties were made with most of the tribes in the 1760s, with official recognition of Maroon autonomy. Their autonomy would continue until after colonial rule.[10] Colonial leprosy politics were not directed at and did not apply to the Native Americans and the Maroons.

Travellers, merchants, soldiers, and slaves would arrive on ships from Europe and Africa at the mouth of the Suriname river. After passing the beach of Braamspunt, they would sail up the river and anchor at Fort Amsterdam. From there, the river led on to the administrative and social centre of the colony, the city of Paramaribo, where most of the Europeans lived. On arrival in Paramaribo in 1773, the Scottish–Dutch mercenary soldier John Gabriel Stedman was pleased to see a town that appeared 'uncommonly neat and pleasing, the shipping extremely beautiful, the adjacent woods adorned with the most luxuriant verdure, the air perfumed with the utmost fragrance, and the whole scene gilded by the rays of an unclouded sun.'[11] However, it was not long before he was confronted with the other side of the tropical climate and the country: health hazards, disease-ridden jungles, and instances of brutal slave treatment.[12]

Racial differences in susceptibility to disease

Most sufferers of boasie in Suriname were Africans and thus Europeans came to the conclusion that boasie had come to the New World from Africa. Ideas about boasie's African connections were related to more general ideas of possible racial differences in disease susceptibility – notions that doctors and surgeons in the Dutch West Indies shared with those from the French and British West Indies. For instance, in 1721, Laurens Horst, a physician and superintendent of the slave depot on the island of Curacao, claimed to have much experience with the 'totally different' character of European and African diseases, about which he planned to write a book.[13]

Others developed similar ideas independently. In 1745, Laurens Storm van 's-Gravesande, the governor of the Dutch colony of

Essequibo (to the west of Suriname and now part of the Republic of Guyana), wrote in a report on an outbreak of smallpox in his fortress: 'I found it very noteworthy that among both my children and others in the colony no one who was born in Europe was so affected [as] the natives of these countries.'[14] Half a century later, the German physician Ernst Karl Rodschied, who practised in Essequibo in the 1790s, found that classical European medicine could not be used in the treatment of African slaves without making modifications. Rodschied, who apparently had only had limited contact with Africans, thought that slaves had few passions, did not mind being slaves, and were really only interested in keeping themselves alive and (if male) in sexual contact with women. According to Rodschied, Africans had brought 'strange' diseases from Africa and their constitution was quite different from that of the Europeans owing to their distinct customs and habits. Moreover, because of the difference in skin colour, 'Hippocratic semiotics', the recognition and interpretation of physical signs and characteristics the European doctors had learned had to be modified. The German physician thought that the passions of the mind (that according to premodern medicine exercised an important influence on health) were radically disparate in Africans and Europeans.[15]

Public hygiene policies, medicine, and slave labour management

The first public health measures against leprosy in Suriname were targeted at the African slaves beginning with an edict issued by the governor in 1761. This edict was essentially a revised version of an edict in 1728. In the 1728 edict, slaves with yaws or other potentially contagious diseases were prohibited from travelling on public roads under penalty of a fine to be paid by their master. Boasie or leprosy was not specifically mentioned at the time.[16] However, the 1761 edict did specifically mention boasie and prohibited slaves with signs of the disease from travelling on public roads.[17] Both edicts had a clear racial and discriminatory nature, since they were only directed against slave movements and not Europeans, even though the 1728 edict was particularly concerned with the danger of potential contamination of white children.[18] The edicts were meant to keep African slaves off the roads, but were not successful since slave owners continued to send afflicted slaves to surgeons in the city of Paramaribo. Thus, a new edict was issued in 1764 prohibiting medical treatment of slaves in the city. Surgeons would have to travel to the plantations to attend to slaves.[19]

These edicts attest to the increase in visibility of leprosy. Rolander was not alone in his observations. Philippe Fermin, a Berlin-born descendant of French Protestant refugees who practised from 1754 to 1762 as a medical doctor in Suriname, spoke of an epidemic. He believed that boasie was highly contagious: *'elle le deviendroit sans les precautions convenables'* ('the disease will be contracted if decent precautions are not taken').[20] According to Fermin, most sufferers were slaves, but there were a few European patients as well.[21] Unfortunately, we do not have more detailed descriptions of these European sufferers, but the edicts of 1761 and 1764 show that Suriname's colonial government had become convinced that boasie was threatening Europeans as well as Africans. The contagiousness of the disease made it a possible threat to slave society and its economy as a whole, rather than only a medical problem.

Health management around boasie was part of a broader attempt by the colonial government to gain more control over healthcare in Suriname. Like all Caribbean colonies, Suriname was continuously plagued by various diseases that were thought to be contagious: all kinds of 'fevers', smallpox, dysenteries, and so forth. The colonial government felt the need for improvements in medical care and in 1766 it put medical practitioners in Suriname under closer scrutiny. These practitioners had to show qualifications or pass an examination before being allowed to set up medical practice. Yet it would take another fifteen years, until 1781, before a board of supervisors, the Collegium Medicum, was instituted. Modelled on the healthcare system of the city of Amsterdam, the Collegium Medicum regulated the activities of doctors, surgeons and pharmacists, and advised the colonial government on health matters.[22]

In the meantime no apparent progress was made in the fight against boasie. This was partly related to the paradoxical status of boasie sufferers. Most of the visible sufferers were Africans who were supposed to be kept out of sight, if not isolated completely from Europeans and healthy slaves. Stedman wrote that in the 1770s, the sufferers were segregated on the plantations.[23] But to their owners these sufferers were also commodities expected to bring in profits, which should be taken literally.

It was not until 1828 that slaves in Suriname were legally regarded as persons instead of commodities. In health issues, as well as other matters, they were completely subordinate to the demands and economics of the slave system.

African slaves were needed to sustain the plantation economy, but they did not bear the plantation working conditions well. In the West Indian colonies, mortality rates were high in the eighteenth century, at least so far as can be ascertained from the existing sources. For instance, on Jamaica where relatively complete records survive, the annual rate of the natural decrease of the slave population (the excess of the crude death rate over the crude birth rate) is estimated at thirty per 1,000 in the first half of the eighteenth century and twenty-five in the 1750s–1770s. Some contemporaries even spoke of annual rates of fifty or sixty per 1,000 in the British colonies.[24] Suriname was no exception. The annual rate of natural decrease here was equally high, and was estimated to be at least forty per 1,000 slaves.[25] At the same time, the import of new slaves was limited. In this context, the passage in Rolander's diary quoted above needs to be reread: 'Blacks infested with this disease [boasie] are automatically relegated to a remote corner of the plantation', he wrote, while adding, 'where they serve as guards and spend the rest of their time alienated from friends to keep the entire servile throng from contracting it'.[26] The plantation owners intended their slaves to continue performing some sort of physical labour, even when afflicted with a horrid disease. Hence, the owners did not openly resist the new public health measures, but rather chose to ignore them. From this perspective, it is significant that the edict of 1761 attempted to encourage slave owners to keep their slaves with leprosy off the public road. Unfortunately, the edict of 1764 showed that this did not have the expected effect. Either the owners disregarded the edict and sent their slaves on the road anyway, or the slaves acted and travelled on their own. The sanctions on these activities – fines for the owners – were not strictly enforced. Private interests prevailed over public-spirited motives since many owners did not wish to sacrifice their interests to those of the slave economy as a whole.

The financial cost of boasie among slaves could be quite high. Slaves were expensive and not always easy to come by, especially when highly qualified. The prices and estimated worth of slaves varied widely. The figures that are available for one plantation vary from 1,000 guilders or more for a highly qualified slave, such as a carpenter or a 'dresiman' or 'dressy negro' (a black surgeon), to 100 guilders for a healthy female house slave.[27] Regardless of the financial value of the slaves, the costs of medical treatment of a black patient by a white surgeon or doctor could significantly increase depending on the number and length of the treatments.[28]

Therefore, slave traders and owners did not hesitate to try to sell off slaves with signs of boasie and to disguise any visible symptoms. For buyers, it was not easy to recognize signs of the disease in its early stages. In 1780, the governor of Suriname issued a new edict explicitly prohibiting the sale of slaves who had leprosy (or mental illness), a step he would not have had to take if this had not been common practice.[29]

Ten years later, in 1790, the colonial government issued even stricter measures suggesting that the incidence of leprosy was not decreasing. Every slave owner was now obligated to report possible leprosy among slaves to the board of medical supervisors, the Collegium Medicum. If the doctors and surgeons of this supervising institution found a slave afflicted with leprosy, the sufferer had to move to a special settlement for those with leprosy.[30] This settlement was called Voorzorg ('Prevention') and was established on an uncultivated bank of the Saramacca River. On 21 December 1791, the first group of seven sufferers of leprosy was deported to this colony.[31] Little is known about Voorzorg. When the first supervisor arrived in 1795, the asylum had about 200 inhabitants. Their number expanded to 300 in 1797, 400 in 1808, and 500 in 1812.[32] This meant that by 1812 almost one in every 100 slaves lived in the asylum. In practice, Voorzorg was nothing more than an isolated place controlled by the military without medical supervision. Shortly after the Voorzorg had been founded, the Dutch ship's doctor, social critic, and an occasional secret agent, Pieter van Woensel visited the settlement. He noted about 140 sufferers, some of them also afflicted with venereal diseases. He left with the impression that the goal of the Voorzorg was to segregate rather than cure the sufferers in an attempt to prevent further contagion of the population.[33]

One other eyewitness account exists dated thirty years after Van Woensel's visit. In 1824, the Dutch Catholic priest Joannes Willemsen went to Voorzorg. With some difficulty, Willemsen received permission to enter the establishment from the commander of the military post situated on the opposite bank of the river from the settlement. However, permission for entry was denied to the priest's Afro-Surinamese servants.[34] Voorzorg seems to have been more or less a colony within the colony of Suriname, and kept isolated by thirty military soldiers. It is most likely that the settlement had a high degree of self-sufficiency. Unlike later leprosy asylums, there was no medical or religious supervision or disciplining of the sufferers.

Not only slaves were sent to Voorzorg since, per the edict of 1790, free Africans and mulattos with leprosy were also placed in the asylum. Europeans with the disease, on the other hand, were allowed to stay in isolation on their plantation or in their city home. It is intriguing that Van Woensel encountered one European in Voorzorg. This was a Dutchman who probably was a poor white without social connections.[35] So sufferers of leprosy who were of 'inferior' race and/or social status were removed to a place of isolation to neutralize their potential threat to the white-dominated slave society. However, they did not receive medical treatment and/or social and religious discipline as would be the case in the nineteenth and twentieth centuries.

An astute observer, Van Woensel was sceptical about the efficacy of the anti-boasie measures. He wrote that the colonial government's attempts to prohibit the sale of slaves without proper medical examination were in vain.[36] In 1791, the colonial government tried to prevent the import of new slaves with boasie. Captains of slave ships had to pay a 100-guilder fine for each slave diagnosed with boasie upon medical examination of the human cargo held on arrival in Suriname. The slave should then have been sent to Voorzorg. But, the edict of 1791 crossed the interests of the powerful commercial companies that ran the slave trade, since it meant an increase in the cost of new slaves from Africa. Thus, in 1792, the fine was revoked.[37]

The interests of the local elite, the plantation and slave owners, did not always coincide with those of the colonial state as represented by the governor appointed by the Society of Suriname that owned the colony. In fact, there was continuous tension between the two parties. The rich planters and merchants of the colony were expected to make financial contributions to the colonial state, but they were seldom content with government policies.[38] In the case of boasie, the physicians' contribution to disease control was hindered by conflict with slave holders' interests. At the same time, the physicians' medical contribution reflected a commonality of fears, prejudices, and interests lying beneath the conflicts between the colonial elite and the colonial government, as is evident in the Schilling's analysis and his framing of a racial pathology.

Godfried Schilling and the colonial and racial framing of leprosy

Medical expertise was a key part of slave labour management, which included managing boasie. Doctors and surgeons had defined the disease in the 1750s as a contagious threat and identified it as leprosy; thus, they had to identify the sufferers of the disease in order to isolate them from society. This, rather than treatment, was their main task. Specialized medical knowledge was needed to accomplish this goal, and Godfried Wilhelm Schilling played a key role.

Little is known about Schilling's early life. He was born in 1733 or 1734 in Wijk bij Duurstede, a town not far from Utrecht in the centre of the Netherlands.[39] Schilling trained as a surgeon before travelling to Suriname. Probably, he was apprenticed to a master surgeon in his youth and, after receiving some sort of certificate of proficiency, went on his *Wanderjahre*, travelling throughout the world and gaining experience before he took a ship to Suriname. We know Schilling went there as a surgeon, although it is not clear if he was ship's surgeon for one of the trading companies (perhaps on a slaver) or was already in the service of the Society of Suriname. In his treatise on leprosy, he refers to his observations in both the 'hot' and the 'cold' regions of America, suggesting that he spent time in both North and South America.[40] After his initial voyage to Suriname, Schilling went back to the Netherlands. He wrote about his observations and thoughts on leprosy in an MD thesis for which he received a medical doctorate at the University of Utrecht in 1769. At the same time, he wrote a treatise on yaws. Both treatises were translated from Latin into Dutch and then published in the Dutch Republic.[41]

It is significant that Schilling's treatises in Latin on diseases that were hardly known in the Dutch Republic were translated into Dutch for more general readership. A primary aim of these translations was to transmit useful knowledge for buying and managing slaves to people working in the slave trade and slave labour management. It is especially significant that Schilling's treatise on yaws, a skin disease that in Suriname was even more widespread among slaves than boasie, was published in the town of Middelburg in the Netherlands. The town, situated in the maritime province of Zeeland, was home to the largest Dutch slave-trading company of the eighteenth century: the MCC (Middelburgsche Commercie Compagnie). The publication of Schilling's treatises in Dutch can be explained by the fact that one of the problems the slave traders and especially the MCC surgeons faced was that when buying slaves in Africa, they were often deceived by slave merchants, who would disguise the symptoms of illness and the age of their human merchandise.[42] Thus knowledge of these 'exotic' diseases was vital within the Dutch Republic.

Through his treatises on leprosy and yaws, Schilling became established as a leading authority on the health and illness of African slaves. Therefore, it is not surprising that in the 1770s he was back in Suriname, probably as a physician at the military hospital in Paramaribo that was opened in 1760.[43] In 1789, Schilling is mentioned as a doctor of medicine at the hospital.[44] According to a traveller called Ludwig, at the end of the 1780s, Schilling was president of the Collegium Medicum, the medical supervisory board of Suriname. He lived opposite the hospital and had an extensive private practice.[45] Ludwig estimated that by the end of the 1780s, the doctor had a yearly income of between 5,000 and 6,000 guilders.[46]

As an 'adventurer-scientist', Schilling turned out to be quite successful. 'Adventurer-scientists' who travelled in the colonies had to possess specific qualities in order to succeed in gathering tropical medical knowledge.[47] Not only were they required to have an eye for useful knowledge and to be capable of making exact observations, but they also had to be able to function under difficult conditions: unusual weather, stress, and a general climate of violence and war. The word 'adventurer' is used in its seventeenth- or eighteenth-century spirit. In eighteenth-century Dutch, an *avonturier* ('adventurer') was a fortune seeker venturing into the unknown. It is exactly because Schilling had the qualities of an adventurer that as a scientist he was better qualified for research in Suriname than someone like Rolander. Schilling's qualities as an adventurer helped him to make the scientific observations that were published in his treatises.

Schilling's framing of leprosy

One of the problems in Schilling's analysis of boasie was the European medical authors who had written about leprosy since antiquity – Celsus, Avicenna, Hildanus, and others. These men had described the symptoms of leprosy in various ways and their terminology was rather confused. Leprosy was called 'lepra arabum', or 'elephantiasis graecorum', while 'lepra graecorum' was used to designate more elephantiasis-like afflictions in the legs.[48] Schilling decided that boasie was some kind of 'melaatschheid' (the Dutch word for leprosy) or lepra arabum because of two characteristic signs: changes in skin colour on affected parts of the body and anaesthesia or insensitivity of those parts.[49] He believed that the disease was contagious, just like European leprosy.

How did this contagion spread? Schilling was convinced that leprosy could not occur without the presence of a 'special substance' or 'a certain poison' that could become virulent when the climate or diet had weakened a person's constitution. Schilling conceived the effect of this leprosy poison in the framework of premodern classical medicine. By eating poorly digested food, the 'chyle' in the body – the substance that according to Galen transformed food into blood – became tough, thick, and sour. The thick, sour chyle was thought to produce thickened blood, which resulted in thickened glands and nerves and stifled physical sensations. The body was then vulnerable or predisposed to the disease. The abnormal thickening of the fluids prevented their healthy evaporation from the body. If the weakened body was then contaminated with a contagious leprosy poison, it contracted the disease.[50] The body could, however, be strong enough to withstand the poison; or, if the first signs of the disease were manifest, one could stop the progress of the disease for ten to twenty years by adopting a healthier diet and lifestyle.[51]

Since Schilling believed that the prevalence of leprosy was greater in Suriname and on the Caribbean islands than in North America, he stressed the importance of dietary and environmental factors in the aetiology of the disease. He believed that in the Caribbean the normal diet, air, and (bad) quality of the drinking water were quite similar to Africa, but different from the colder regions of the north. In the hot climates, the diet – which included fat, rotten meat and fish, and bad water – caused the disease. The hot air aggravated problems with the evaporation of bodily fluids. The leprosy poison spread through physical contact, that is, by sexual intercourse or by contact with the ulcer and wound exudate.[52] In his medical practice, Schilling ordered the clothes of those people who were suspected to have died from boasie to be burned whenever possible to prevent further contagion.[53]

Schilling's ideas were not innovative or unique when considered within medicine at the time. In the eighteenth century, European physicians considered leprosy in particular from the perspective of humoral pathology. However, unlike his colleagues in Europe, Schilling added moral and racial elements to his medical views by emphasizing the role of the African slave in the transmission of leprosy.[54] These elements show how his framing of the disease was connected to the context of the slave society, and was driven by the interests of the slave economy.

In emphatically pointing out that Africans were the carriers of the disease, Schilling justified the edicts of 1780, 1790, and 1791, which attempted to control the slave trade and isolate sufferers. Schilling felt that whenever Europeans had physical contact with Africans, they were in danger of the disease's contagion. Europeans were especially at risk if they did not adopt a healthy diet and adjust to the hot climate. Measures had to be taken to prevent Europeans' contact with sufferers of leprosy. There were distinct moral connotations inherent in this view. Africans with leprosy had contracted the disease because of their susceptibility; they lacked the self-control and level of civilization needed to withstand the leprosy poison. Europeans were at risk in Surinamese society because of close contact between slave holders and slaves. This risk was not only physical, but also moral. At first glance, it seems that Schilling put the blame entirely on the Africans. However, the moral nature of his leprosy diagnosis did not spare the resident Europeans either. This was not because Europeans had instituted the slave trade and were responsible for bringing Africans and leprosy to America (a critical opinion put forward by Van Woensel). Rather, it was because Schilling felt that Europeans had difficulty controlling their sexual urges.[55] According to Schilling, the hot air of the tropics made Europeans lecherous. In North America, where the air was not as hot, and where there were relatively more European women and fewer slave women than in Suriname, intercourse between white men and African women did not occur as frequently. Schilling claimed that when this intercourse occurred in North America it was owing more to European curiosity than to 'necessity'.

According to Schilling, Africans were also beset by lewdness (Dutch: 'geilheid'), and he argued that the more anyone in Suriname was ruled by lust, the more leprosy contagion would continue to spread from slave to master.[56] Van Woensel agreed with Schilling in this regard. According to him, it was customary for white men to have black concubines and these women had given the Europeans the legacy of leprosy.[57]

In Schilling's view, unhealthy and unclean living stimulated the spread of leprosy. The European male became threatened by leprosy the moment he lost his self-control and stooped to the level of a race that was held to be inferior. In this way, medical, social, and moral degeneration went together and endangered European dominance in a society of largely Africans. Leprosy became a threat to the empire. African slaves were thought to have brought the disease across the Atlantic to the Caribbean, where Europeans were then infected. Europeans could then in turn bring leprosy back to the Netherlands, where it had become extinct. These fears seemed to be justified by incidences of leprosy among Europeans in Suriname. In 1786, the Lutheran Council of Paramaribo requested that the government support the Europeans who were suffering from boasie and had turned to the Lutheran poor relief for money and accommodation.[58] Four years later, Schilling had the clothes burned of a deceased suspected leprosy sufferer with whom he had worked closely, the military surgeon-major Heijder.[59] Van Woensel expressed fears that contagion of Europeans by Africans would bring leprosy back to the Netherlands, transmitted by returned travellers to fellow-Europeans.[60]

Diagnosis and treatment

The racial and sexual aspects of leprosy made it problematic to control the disease. Therefore, Schilling suggested that it was essential to make a thorough medical examination of slaves before they were allowed into owners' houses.[61] However, early detection of the disease alone proved to be difficult. Fermin had found it problematic to recognize the disease in an early phase, claiming that the symptoms (the discoloured spots) were very similar to ringworm (a fungal infection of the skin).[62] To Schilling, the issue was further complicated as the semiotics of European medicine had to be differentiated according to race. The spots that were an early sign of leprosy were either red tending to a rather pallid colour, or white tending to blue, yellow or red. In white people, white spots were hard to discern; in black people the same held for red spots.

Apart from this, there was the problem of the stigma attached to boasie or leprosy. Schilling found that often people with signs of boasie would not go to a physician, out of either ignorance or shame, especially Europeans. They would hide the signs under their clothes or pretend to feel pain despite their anaesthesia when undergoing medical examinations. Therefore, Schilling advised doctors and surgeons to do an examination when the patient was asleep, putting a needle or a knife in the suspect spots or even scorching them with fire.[63] He might have learned the ploy with the needle or knife from Fermin, whose treatise on Surinamese diseases had been published in 1764. Fermin claimed to have learned the trick from an 'old Negress', showing that European doctors did try to make use of African ideas and practices.[64] They also tried to find out about African therapeutic methods. Slave holders' medical knowledge to some extent made use of slave medical knowledge. One of the problems for the European doctors was how to obtain this knowledge, which was hard to achieve, since slaves were not open towards their masters.[65]

The basic principle behind Schilling's therapeutic recommendations was for the physician to assist nature in curing the disease by prescribing a healthy lifestyle to his patients. Fermin was of the opinion that boasie was 'absolutely incurable'.[66] Schilling thought otherwise. According to him, the first step in the therapeutic process was the prescription of a diet of bread, vegetables, and juices, with only a sparing use of dairy products. When the patient's body had been strengthened through this diet, the physician could then try to remove leprosy poison from the body. He would need to purge the body by prescribing lukewarm baths, fumes, and walking exercise. Schilling advised that physicians should administer sarsaparilla and kina to make the body sweat and treat the skin with balsams. In this way, Schilling claimed the disease could be cured in half a year.[67] Other observers in Suriname were less optimistic. The plantation manager Anthony Blom observed in the 1780s that the body parts of his leprous slaves rotted and died. According to Blom, physicians did not have a cure.[68] Van Woensel was sceptical of Schilling's treatment, although he was convinced that sweating was the best way of treating the disease. According to him, plantation slaves who worked in heated attics to dry coffee beans and who sweated continuously were cured of skin diseases.[69] Other observers differentiated between several types of leprosy. It was believed that 'dry' boasie could be healed, but not 'wet' boasie.[70]

Treatments on the plantations

Most slaves, of course, did not receive the extensive treatments that Schilling had prescribed. On the plantations, they were segregated in separate huts and they generally continued to work until the signs of the disease became too severe. They were then left to their fate.[71] African healers, who were important providers of slave healthcare on the plantations, often administered treatment, since there was a lack of trained doctors. On the one hand, Schilling found it beneath his dignity to investigate these healers' methods – or at least he claimed so. On the other hand, he was curious. He gave one example of what these so-called 'uncivilized people' could achieve with 'a superficial and erroneous medicine'. Schilling bought a 'secret cure for leprosy' from a female African healer. Her method was not unlike Schilling's own. She purged her patients with plant medicine, had them exercise, treated their spots with herbs, and had them healed in three or four months.[72] However, for this African healer, as well as European physicians and surgeons, diagnosing boasie or leprosy must often have been difficult and unclear, and leprosy often confused with other skin or venereal diseases.

Conclusion

Driven by the needs and interests of a slave society and an economy that felt threatened by leprosy, Schilling framed the disease as tainted with negative connotations, originating in Africa, caused by unhealthy living conditions, and related to disreputable sexual morals. He adapted classical medical concepts, such as the influence of passion, weather, or diet on health and disease, to the physical and social conditions of the Caribbean slave societies and developed a racial pathology of leprosy. In this pathology, medical, moral, and political aspects were intertwined. Africans were seen as the source of the danger of contagion, had to be controlled, and when necessary isolated. Europeans had to exercise control over both their subjects and themselves. Contracting boasie was not purely a physiological process. As was common in disease perception in premodern medicine, the 'passions of the mind' were of decisive importance. These passions were the precise difference between superior and inferior and between slave owner and slave. In short, the consequence of Schilling's pathology was that boasie's contagion was a threat to the difference between master and servant and undermined the very essence and foundation of slave society. Adaption of classical medicine to tropical slave societies also offered promising perspectives on prevention and treatment. Keeping the Europeans physically and morally strong and African sufferers isolated held contagion at bay, and disease could then be conquered by returning strength to the diseased by purging, exercise, and a correct lifestyle.

By the end of the eighteenth century, something remarkable and new occurred for leprosy in Suriname. Leprosy was framed as more than a purely medical problem. It was also a problem of geography and race. The hot climate stimulated the expression of the disease in those with a fatal predisposition. Furthermore, leprosy was an economic problem and a threat to the functioning of the slave economy. Finally, leprosy was a social and political threat. The disease was framed as a danger to European dominance. The sufferers of the disease who threatened this dominance had an inferior racial and/or social status. By the end of the century, the solution was to segregate and isolate them, and leave them to their fate.

Notes

1 D. Rolander, 'Journal', in L. Hansen, D. Goodall, and J. Dobreff (eds.), *The Linnaeus Apostles: Global Science and Adventure*, vol. 3, bd. 3 (London: IK Foundation, 2008), pp. 1217–564, on p. 1339.
2 Rolander, 'Journal', p. 1483.
3 Boasie is the name of a town in present-day Ghana, about 30 miles to the north of the capital Accra. A Dutch dictionary of 1855 would claim that boasie was the name given in Angola to an ulceration that looked like elephantiasis. See L. C. E. E. Fock, *Natuur- en geneeskundig etymologisch woordenboek* (n.p.: J. Noorduyn, 1855), p. 170. Other meanings of 'boasie' are, according to twentieth-century Dutch dermatologist Robert Simons, 'impure' and

'under a rock'. See R. D. G. P. Simons, *Lepra. De lepra-bestrijding in Suriname en de noodzakelijkheid harer reorganisatie* (Amsterdam: Scheltema and Holkema, 1950), p. 10; R. D. G. P. Simons, *Bijgeloof en lepra in de Atlantische Negerzônes* (Paramaribo: Radhakishun, 1959), p. 33.

4 Here 'leprosy' is used to refer to 'boasie', 'melaatschheid' and 'elephantiasis graecorum' interchangeably, as these were understood by eighteenth century physicians, and not to the current medical understanding of leprosy or Hansen's disease.

5 On premodern and early modern medical thought on leprosy: L. Demaitre, *Leprosy in Premodern Medicine: A Malady of the Whole Body* (Baltimore, MD: The Johns Hopkins University Press, 2007).

6 Rolander, 'Journal', p. 1483.

7 The name Schilling is sometimes spelled as 'Schelling', but he signed his own name as 'Schilling': C. H. Landré, 'Naschrift bij P. Duchassaing, Over de Elephantiasis Arabum in West-Indië', *West-Indië. Bijdragen tot de bevordering van de kennis der Nederlandsch West-Indische kolonien*, 2 (1858), pp. 222–33, on p. 224. In a previous publication I presented Schilling as a Prussian (S. Snelders, 'Leprosy and slavery in Suriname: Godfried Schilling and the framing of a racial pathology in the eighteenth century', *Social History of Medicine* 26 (2013), pp. 432–50). Natalie Zemon Davis has since then discovered that Schilling was a Dutchman, born in 1733 or 1734 in Wijk bij Duurstede: email communication to the author 29 June 2015; N. Z. Davis, 'Physicians, healers, and their remedies in colonial Suriname', *Canadian Bulletin of Medical History/Bulletin canadien d'histoire de la médecine* 33 (2016), 3–34; wedding register of the Reformed Church of Paramaribo: National Archive, The Hague, Doop-, trouw- en begraafboeken van Suriname (herafter DTB), Nr. 1, 323, 12 May 1764, scan 271.

8 G. W. Schilling, 'Dissertatio medica inauguralis de lepra' (MD thesis, University of Utrecht, 1769); G. W. Schilling, *Verhandeling over de melaatsheid* (Utrecht: J.C. ten Bosch, 1771); G. W. Schilling, 'Animadversiones in Ouseelianam et additamenta ad suam de lepra dissertationem', in J. D. Hahn (ed.), *De lepra commentationes* (Leiden: Abr. van Paddenburg, 1778), pp. 119–203.

9 For the sake of consistency and clarity, I will here use 'Europeans' for all whites, including those born in Suriname (who were called 'Creoles' at the time, while Europeans in general were also referred to as 'blanken' or whites). To avoid confusion I will use the word 'Africans' or 'Afro-Surinamese' for slaves and freed people of African descent. Population figures: R. O.

Beeldsnijder, '"Om werk van jullie te hebben". Plantageslaven in Suriname, 1730–1750' (PhD thesis, Leiden University, 1994), pp. 264–6; B. van der Oudermeulen, 'Iets tot voordeel der deelgenooten van de Oost-Indische Compagnie en tot nut van ieder ingezetenen van dit gemenebest kan strekken', in D. van Hogendorp (ed.), *Stukken, raakende den tegenwoordige toestand der Bataafsche bezittingen in Oost-Indië en de handel op derzelve* (The Hague: J.C. Leeuwesteyn, 1801), pp. 327–38, on pp. 327–8. Beeldsnijder is probably right to assume that '1738' in the memoir of Van der Oudermeulen is a typographical error and should be read as '1783'.
10 B. P. C. Scholtens, *Bosnegers en overheid in Suriname. De ontwikkeling van een politieke verhouding 1651–1992* (Paramaribo: Afdeling Cultuurstudies, 1994).
11 J. G. Stedman, *Narrative of a Five Years' Expedition Against the Revolted Negroes of Surinam, in Guiana on the Wild Coast of South America from the Years 1772 to 1777* (Amherst, MA: University of Massachusetts Press, 1972), p. 14.
12 S. Snelders, *Vrijbuiters van de heelkunde. Op zoek naar medische kennis in de tropen 1600–1800* (Amsterdam: Atlas, 2012), pp. 15–17.
13 National Archive, The Hague, Tweede West-Indische Compagnie Archive (hereafter NA 2WC), Nr. 576, Laurens Horst to West India Company, February 1721.
14 National Archives, Kew, Colonial Record Office (hereafter CRO), CRO 116/29, Laurens Storm van 's-Gravesande to the Chamber of Zealand of the West India Company, February 1745.
15 E. K. Rodschied, *Medizinische und Chirurgische Bemerkungen über das Klima, die Lebensweise und Krankheiten der Einwohner der Holländischen Kolonie Rio Essequibo* (Frankfurt: Jaegerschen Buchhandlung, 1796), pp. 126–38.
16 'Plakkaat, Bescherming tegen besmettelijke ziekten', 30 November 1728, in J. T. de Smidt (ed.), *Plakkaten, ordonnantiën en andere wetten, uitgevaardigd in Suriname 1667–1816* (Amsterdam: Emmering, 1973), pp. 395–6.
17 'Plakkaat, Voorschriften in verband met besmettelijke ziekten van slaven', 4 February 1761, in De Smidt (ed.), *Plakkaten*, pp. 707–8.
18 'Plakkaat', in De Smidt (ed.), *Plakkaten*, pp. 395–6.
19 'Notifikatie, Het verplegen van slaven in chirurgijnswinkels', 22 February 1764, in De Smidt (ed.), *Plakkaten*, pp. 780–1.
20 P. Fermin, *Traité des maladies les plus frequentes à Surinam, et des remedes les plus propres à les guérir* (Maastricht: Jacques Lekens, 1764), p. 127.

21 Fermin, *Traité des maladies*, p. 127.
22 G.-J. Hallewas, 'De gezondheidszorg in Suriname' (PhD thesis, Groningen University, 1981), p. 55.
23 J. G. Stedman, *Narrative of a Five Years' Expedition against the Revolted Negroes of Surinam, in Guiana and the Wild Coast of South America from the Year 1772 to 1777*, vol. 2 (London: J. Johnson & J. Edwards, 1796), p. 275.
24 J. R. Ward, *British West Indian Slavery, 1750–1834* (Oxford: Clarendon Press, 1988), p. 123.
25 H. Buddingh', *De geschiedenis van Suriname* (Amsterdam: Nieuw Amsterdam, 2012), p. 88. On the reproduction of slaves and the scarcity of slaves in eighteenth-century Suriname: A. A. van Stipriaan Luïscius, 'Surinaams contrast. Roofbouw en overleven in een Caraïbische plantage-economie' (PhD thesis, Vrije University Amsterdam, 1991), p. 213.
26 Rolander, 'Journal', p. 1339.
27 G. Oostindie, *Roosenburg en Mon Bijou. Twee Surinaamse plantages, 1720–1870* (Dordrecht: Fortis, 1989), pp. 106–7.
28 J. F. Ludwig, *Neueste Nachrichten von Surinam* (Jena: Akademischen Buchhandlung, 1789), p. 165.
29 'Notifikatie, Zieke of krankzinnige slaven mogen niet worden verkocht', 10 February 1780, in De Smidt (ed.), *Plakkaten*, pp. 971–2.
30 'Plakkaat, Voorzorgsmaatregelen tegen melaatsheid', 28 May/4 June 1790, in De Smidt (ed.), *Plakkaten*, pp. 1144–7.
31 A. Bosser, *Beknopte geschiedenis der katholieke missie in Suriname* (Gulpen: Alberts, 1884), pp. 177–8; 'Report Commissie van Geneeskundig Onderzoek en Toevoorzigt', *Nieuw Praktisch Tijdschrift voor de Geneeskunde in al haren omvang*, 1 (1849), pp. 554–65, on p. 556.
32 'Report Commissie van Geneeskundig Onderzoek en Toevoorzigt', p. 556.
33 P. van Woensel, 'West-Indische fragmenten', in A. Hanou (ed.), *De lantaarn* (Amsterdam: Athenaeum-Polak & Van Gennep, 2002), p. 52.
34 Bosser, *Beknopte geschiedenis*, pp. 170–2.
35 Van Woensel, 'West-Indische fragmenten', p. 52.
36 Van Woensel, 'West-Indische fragmenten', p. 51.
37 'Publikatie, Voorzorgsmaatregelen tegen melaatsheid', 20/27 May 1791, in De Smidt (ed.), *Plakkaten*, pp. 1159–60; 'Publikatie, Voorzorgsmaatregelen tegen melaatsheid', 29 February 1792, in De Smidt (ed.), *Plakkaten*, pp. 1167–8.
38 See Buddingh', *Geschiedenis*, pp. 35–50; J. M. W. Schalkwijk, *The Colonial State in the Caribbean: Structural Analysis and Changing Elite Networks in Suriname, 1650–1920* (The Hague: Amrit, 2011).

39 DTB, Nr. 1, 323, 12 May 1764, scan 271.
40 Schilling, *Verhandeling over de melaatsheid*, p. 1.
41 Schilling, 'Dissertatio medica inauguralis de lepra'; G. W. Schilling, *Diatribe de morbo in Europa pene ignoto, quem America vocant Jaws* (Utrecht: J.C. ten Bosch, 1770); G. W. Schilling, *Geneeskundige verhandeling van eene in Europa byna onbekende ziekte, bij de Amerikanen JAWS genoemd* (Middelburg: Christiaan Bohemer, 1770); G. W. Schilling, *Verhandeling over de melaatsheid*.
42 This problem is discussed in D. H. Gallandat, *Noodige onderrigtingen voor de slaafhandelaren* (Middelburg: Pieter Gilissen, 1769). Gallandat, a Swiss surgeon who had served on slavers of the MCC, published this study around the same time as Schilling's treatises were published.
43 J. Karbaat, '200 jaar militair hospitaal in Paramaribo (1760–1960)', *Nederlands Militair Geneeskundig Tijdschrift* 13 (1960), pp. 355–64; E. van der Kuyp, 'Surinaamse medische en paramedische kroniek 1494–1899', *Surinaams Medisch Bulletin* 9 (1985), pp. 1–67, on p. 18.
44 *Surinaamsche Almanach, op het jaar onzer Heere Jesu Christi Anno 1789* (Paramaribo: W. H. Poppelman, 1789), p. 21.
45 Ludwig, *Neueste Nachrichten*, p. 164.
46 Ludwig, *Neueste Nachrichten*, p. 165, claims that this was the normal income of a surgeon in Suriname. Schilling would have earned at least that amount, probably more.
47 Snelders, *Vrijbuiters van de heelkunde*; S. Snelders, '"Kapers van kennis". De rol van een boekaniersgeleerde in de circulatie van kennis over ziekten en geneesmiddelen in de tropen', *Studium* 2 (2009), pp. 55–64.
48 On writings on leprosy by other medical authors: Schilling, *Verhandeling*, p. 4; Demaitre, *Leprosy in Premodern Medicine*.
49 Schilling, *Verhandeling*, p. 8.
50 Schilling, *Verhandeling*, pp. 32–40.
51 Schilling, *Verhandeling*, pp. 15–17.
52 Schilling, *Verhandeling*, pp. 26–7.
53 C. L. Drognat Landré, *De besmettelijkheid der lepra arabum, bewezen door de geschiedenis dezer ziekte in Suriname* (Utrecht: J. L. Beijers, 1867), p. 13.
54 On leprosy in eighteenth-century European medicine: Demaitre, *Leprosy in Premodern Medicine*, pp. 132–59.
55 'A poet would say: to avenge themselves for the injustice done to them by us, the Negroes – or rather the Surinamese – have given us this legacy [of leprosy] [my translation]'. Van Woensel, 'West-Indische fragmenten', p. 51.

56 Schilling, *Verhandeling*, pp. 29–32. On sexual relations in Suriname between Europeans and Africans: van Stipriaan, 'Surinaams contrast', p. 396; Buddingh', *Geschiedenis*, pp. 65–9. On other Caribbean colonies: R. Hyam, *Empire and Sexuality: The British Experience* (Manchester: Manchester University Press, 1990), pp. 92–3: 'In the eighteenth century, the West Indies do seem to have been a kind of sexual paradise for young European men: it was almost customary for white men of every social rank (but equally of the lower classes) to sleep with black women.'; K. K. Weaver, *Medical Revolutionaries: The Enslaved Healers of Eighteenth-Century Saint Domingue* (Urbana, IL: University of Illinois Press, 2006), p. 18: 'According to many Enlightenment thinkers, the heat of the tropical sun intensified the sexual longings of the region's inhabitants.'; P. Pluchon, *Nègres et Juifs au XVIIIe siècle. Le racism au siècle des Lumières* (Paris: Tallandier, 1984), p. 214.
57 Van Woensel, 'West-Indische fragmenten', p. 51.
58 H. D. Benjamin, 'Treef en lepra in Suriname', *West-Indische Gids* 11 (1930), pp. 187–218, on p. 211.
59 Drognat Landré, *Besmettelijkheid*, p. 13n.
60 Van Woensel, 'West-Indische fragmenten', pp. 51–2.
61 Schilling, *Verhandeling*, p. 79.
62 Fermin, *Traité des maladies*, p. 128.
63 Schilling, *Verhandeling*, p. 11.
64 Fermin, *Traité des maladies*, p. 128.
65 Snelders, *Vrijbuiters van de heelkunde*, pp. 166–82.
66 Fermin, *Traité des maladies*, p. 127.
67 Schilling, *Verhandeling*, pp. 53–73.
68 A. Blom, *Verhandeling over de landbouw in de colonie Suriname* (Amsterdam: J.W. Smit, 1787), pp. 341–3.
69 Van Woensel, 'West-Indische fragmenten', pp. 52–3.
70 J. J. Hartsinck, *Beschrijving van Guiana of de Wilde Kust, in Zuid-Amerika*, 2 vols. (Amsterdam: Gerrit Tielenburg, 1770), vol. 2, pp. 914–15.
71 Stedman, *Narrative* (edition 1796), II, p. 275.
72 Schilling, *Verhandeling*, pp. 73–8.

2
A policy of 'Great Confinement', 1815–1863

Michel Foucault used the concept of 'Great Confinement' in *Madness and Civilization* to designate the confinement of 'madmen' in seventeenth-century Paris in 'enormous houses'. These houses, he claimed, were not medical establishments; rather they were 'an instance of order, of the monarchical and bourgeois order being organized in France during this period'. According to Foucault, 'More than one out of every 100 inhabitants of the city of Paris found themselves confined there.'[1]

In Suriname, the exclusion of leprosy sufferers that had started in the second half of the eighteenth century began to resemble a 'Great Confinement' in the period from 1830 to 1860. Close to one out of every 100 inhabitants was condemned or suspected of having leprosy or elephantiasis and confined to the Batavia leprosy asylum, or segregated at home or elsewhere. The leprosy asylum did not function in the first place as a medical establishment, but rather as an instance of colonial order. From this perspective, leprosy policy in Suriname in this period can be designated as a policy of 'Great Confinement'.

This chapter investigates the development and implementation of segregation policies after Dutch rule returned to the colony in 1816. The Leprosy Edict of 1830 inaugurated a period of 'Great Confinement' following fifteen years of reorientation and a change in a slave's legal status from commodity to person. The confinement was not limited to leprosy sufferers. It also applied to sufferers of elephantiasis, a 'new' colonial disease often confused with leprosy, which also posed a threat to Suriname. Fears of leprosy and elephantiasis became conflated in the medical and public eye, thus contributing to increased vigilance.

Although the 'Great Confinement' of the nineteenth-century colonial state in Suriname was a continuation of the leprosy politics of the eighteenth-century colonial regime, it was adapted to a new legal situation in which slaves had the status of persons rather than the possessions of their owners. Segregation policies continued on the basis of a racial and sexualized pathology of leprosy, reflecting fears about a loss of white dominance and the threat of the black 'other'. Slaves and the poor were sent to an isolated asylum in the jungle, while better-off whites remained at home. Thus, segregation policies were as much a matter of labour management as public health. However, just as in the eighteenth century, slave owners' private economic interests were not necessarily in accordance with public policies designed to defend the slave economy as a whole. In the messy realities of everyday life, segregation policies could not always be strictly enforced. Confinement policies were intense, especially in the period from 1830 to 1855, but the degree of effectiveness or the thoroughness of the segregation of all leprosy and elephantiasis sufferers remained unclear. Ultimately, it was not the difficulty of enforcing the policy of segregation or doubts about its effectiveness, but rather the approaching abolition of slavery that put an end to the 'Great Confinement'.

Return of Dutch colonial rule and renewed fears about leprosy

In 1795, the revolutionary French armies conquered the Dutch Republic and installed a new government based on the principles of the Revolution and the alliance with (or rather dependence on) France. The Revolutionary and Napoleonic wars had significant consequences for the situation in the Dutch colonies, including Suriname. In 1795, the new Batavian Republic (as the revolutionary Dutch regime was called) dismantled the Society of Suriname and took over direct control of the colony. However, communication and movement between the Netherlands and the colonies were extremely difficult due to British control of the sea. From 1799 to 1802, and from 1804 to 1816, Britain took control of Suriname.[2] British sea power prevented the importation of new slaves after the prohibition of the slave trade by the British Parliament in 1808. However, this prohibition did not affect the essential character of the Surinamese slave society, the plantation economy, nor internal slave trade in the colony.

After Napoleon's defeat in 1815 a new Kingdom of the Netherlands was created and the Dutch governor, Willem van Panhuys, formally took over Suriname from the British in February 1816. He arrived in Suriname accompanied by two military surgeons who would play an important role in how leprosy was framed and managed: Frederik Kuhn and Andries van Hasselaar. They also met with a new disease, which according to them was one of the most lasting legacies of the British regime.[3] In Suriname, a popular notion existed that British slaves had brought this disease from Barbados in 1799.[4] In the British Caribbean, the disease was known as Barbados leg because it was first observed on Barbados in 1726.[5] The Dutch called the disease elephantiasis. In Suriname, it was (and is) known as 'roos'.[6] This was not elephantiasis graecorum or leprosy that had been known since the 1750s, but rather 'elephantiasis arabum' or 'lepra graecorum' in the terminology of classical medical authors. (The adjectives 'graecorum' and 'arabum' mean Greek and Arabian, respectively.) Elephantiasis is characterized by swellings and infections of the limbs and the scrotum, and is accompanied by fevers. The disease has a potentially lethal outcome in the later phases.[7] At first, it was unclear to the Dutch whether elephantiasis arabum was a form of leprosy (sometimes it was called local lepra arabum), or another disease altogether. Was it contagious, and should similar health polices be adopted as for leprosy?

By 1820, a differentiation between leprosy and elephantiasis became more common in Dutch medical literature.[8] The German physician Constantin Hering, who practised in Suriname from 1826 to 1832, made a clear distinction between the two diseases. According to Hering, it was rare that a person contracted both diseases.[9] Even so, at that time, the diseases were thought to have the same aetiology. Elephantiasis was understood to be contagious, but it was believed that a weakened constitution, predisposition by heredity and/or lifestyle (Dutch: 'leefregel'), the impact of the hot climate, and psychological factors (such as sorrow and stress) made one vulnerable to the disease.[10] Once again, the Europeans considered Africans to be particularly at risk since they supposedly overstuffed themselves with unhealthy food, were slow to understand their health situation, and were dirty and unhygienic.[11] On the contrary, the more robust Amerindians living outside civilization in natural conditions supposedly never got the disease. Elephantiasis became such a health problem in Suriname that it was incorporated into the new leprosy edict of 1830, which replaced earlier edicts from the eighteenth century.

At this time, leprosy and elephantiasis were not the most urgent health problems in Suriname since just as in the eighteenth century epidemics were rampant in the colony. In 1819, there was an epidemic of smallpox resulting in 10,000 deaths, mostly slaves, and, in 1823, an epidemic of influenza. An important part of Dutch surgeons' work was to control these epidemics. However, fears of leprosy remained as great as they had been in the eighteenth century. When Governor Van Panhuys arrived in Suriname, he immediately became concerned about an increase of boasie among Europeans. Van Panhuys was himself a plantation owner and had been in Suriname before 1816. He advocated a new idea: racial segregation of boasie sufferers. Van Panhuys wanted separate asylums for black leprosy sufferers and for white sufferers who could not be treated at home. He wanted one asylum for the 'High-Dutch' or to be more exact all non-Jewish European leprosy sufferers, and one for Jewish sufferers.[12] Van Panhuys did not like Jews. In January 1815 he wrote to the Dutch government that there were too many Jews in Suriname, and that they were involved in usury and smuggling. Immediately after his arrival as governor, Van Panhuys again complained about the large number of Jews.[13] Van Panhuys was not alone in his dislike of the Jews. As early as the second half of the eighteenth century, there had been Paramaribo citizens who even wanted to build a Jewish ghetto.[14] In 1812, there was unrest in Paramaribo caused by a Jew named J. C. de Mattos, who had been diagnosed with leprosy and ordered to leave town.[15] Ultimately, the idea of a separate Jewish asylum was quietly dropped after Van Panhuys' early demise from a tropical fever in July 1816. The Jews became increasingly assimilated in society and were even appointed to important government posts.[16] However, Van Panhuys' suggestion of racial segregation for sufferers demonstrated that the returned Dutch colonial rulers were just as afraid of the dangers of leprosy as their eighteenth-century predecessors had been and were looking for an adequate response.

European medical professionals

Where could leprosy sufferers turn for medical care in 1816? What was the Surinamese medical market's supply side? The Society of

Suriname's regulations for colonial medical professionals were still in place in 1816 and obligated these practitioners to show a qualification or pass an examination, or in the case of surgeons, to be 'known' to the Collegium Medicum. The requirements for medical professionals were modernized over the next twenty-five years. In 1824 a new regulation called for the possession of a diploma. In Suriname, revised regulations in 1838 required that medical practitioners' diplomas be obtained at a Dutch university or academy (Dutch: 'hogeschool'); in addition, the candidate had to have the right of citizenship in the colony. However, a surgeon could be educated and trained in Suriname, where an apprenticeship to become a master took five years.[17]

Studies and accounts of the Surinamese healthcare system have focused on Western medical practitioners, institutions, and practices and their contemporary successors.[18] The supply side of the medical market shows greater differentiation. As late as 1832, the Collegium Medicum had to warn pharmacists that they were not allowed to fill prescriptions from non-recognized doctors – a sure sign that this was common practice at the time.[19] Furthermore, there were white plantation owners practising their own forms of treatment, black plantation surgeons or 'dresimen', and black healers operating outside white control. Three sectors overlapped, white 'professional' healthcare professionals, Afro-Surinamese 'folk' healthcare practitioners, and the 'popular' sector of individual patients and their families, social networks, and communities.[20] All healers, whether from the professional or the folk sector, were regarded and referred to as 'dokters' (doctors) by the Afro-Surinamese.[21]

There were only a small number of European surgeons in the colony. For instance, in 1819 there were six in the interior near the rivers and plantations, and eight in Paramaribo, apart from the military surgeons who cared for the soldiers, officers and their families. Two of the surgeons in the city had a medical doctorate.[22] One was Frederik Kuhn, the chief military surgeon and a member of the Collegium Medicum.[23] His view of his colleagues' abilities was quite negative. He believed that the surgeons in Suriname only had superficial medical knowledge and had learned their profession in daily practice so they did not know what to do in serious cases. One might dismiss Kuhn's opinion as coming from a medical doctor looking down upon surgeons as tradesmen with little education. However, if Kuhn is believed, the Surinamese population shared his low regard for surgeons, especially because surgeons sometimes earned money on the side by dealing in liquor and were believed to be poor judges of men.[24] When Kuhn's views were published in 1828, the year of his return to and death in the Netherlands, one surgeon in Suriname was so offended by Kuhn's descriptions of slave treatment in the Paramaribo hospital that he paid for an advertisement in a Surinamese newspaper protesting against Kuhn's views.[25]

In large parts of Suriname, surgeons were not available.[26] On plantations, slaves were dependent on the medical skills of plantation directors (who according to Kuhn also had little knowledge), and the Dress Negroes or dresimen. These were black surgeons who often had been apprenticed for five to six years to white surgeons and Kuhn had a low view of them as well. For Kuhn, 'there is no more merciless creature than a Negro towards his equals, especially a Dress Negro'.[27] Finally, there were female black healers, the Dress Mamas, including those who were prophetesses and herbal healers. A white surgeon had to tread very carefully around these Dress Mamas, or risk losing credit with the black population.[28]

Surgeons learning their trade in practice might not have been a disadvantage in Suriname, since medicine is a skill that has to be learned in practice even today. However, it is likely that the local surgeons were not capable in detecting signs of lepra arabum, managing the disease, or distinguishing lepra arabum from lepra graecorum or other skin diseases. The few surgeons who did have greater knowledge and had received their medical doctorates (such as Schilling and Kuhn), but who lacked experience, might not have necessarily been more capable of managing leprosy and elephantiasis than their more poorly educated colleagues. They did, however, have more influence in shaping public views on leprosy and determining health policies in the colony, thus playing an instrumental role in generating alarm about leprosy during the period of slavery.

Medical perspectives on leprosy and elephantiasis

Schilling's views had set the tone for understanding leprosy. The roles of racial dispositions and sexual behaviour in this understanding return again and again in the writings and judgements of nineteenth-century doctors and others in Suriname. For instance, in 1804, F. van Heshuyzen, a Dutch councillor and financial administrator of the

British government in Suriname, explained to his new bosses that leprosy was contagious and white immigrants were especially at risk when they slept with African or mulatto women.[29] Kuhn was convinced, as his eighteenth-century predecessors had been, that diseases and their expression were differentiated according to race. He claimed that the Africans had their own 'special' diseases, including leprosy, yaws, and dirt eating. Their disease management was supposedly characterized by the prejudices and superstitions that were typical of 'uncivilized' people.[30] Similar racial considerations permeated Kuhn's reflections on elephantiasis. In Paramaribo, Kuhn noticed a greater prevalence of elephantiasis among Africans and mulattos than Europeans. He concluded that it was a constitutional disease that 'came into' the lymphatic system because of hereditary (racial) predisposition and contagion. However, for Kuhn the 'true cause' of the expression of the disease was an inadequate diet combined with an 'onreine en liederlijke levenswijze' (impure and dissolute lifestyle). A diagnosis of elephantiasis and leprosy could imply moral judgement of the sufferer. This judgement was connected to racial elements, since Africans, mulattoes, and Jews were the most afflicted. Kuhn's medical writings were suffused with notions of African inferiority. He wrote that their diet consisted of mealy and not easy digestible foodstuffs: bananas, cassava, yam, unripe fruits, salted meat, and fish. From their early youth, Africans were used to stuffing themselves with this unhealthy food. The slaves were described as dull and dirty. In Kuhn's view, slaves were unable to handle themselves and needed their masters' guidance.[31]

Leprosy and elephantiasis threatened the Dutch colony, and according to Kuhn, it was up to the colonial rulers to impose strict measures on their slaves to fight these diseases. This was ultimately the key message of his writings, as it had been for Schilling on leprosy. The message stimulated as well as justified a stricter segregation policy that was reflected in the most detailed medical study of leprosy in Suriname in the age of slavery published by Andries van Hasselaar (1782–1838). Van Hasselaar was another surgeon who had come to the colony in Van Panhuys' footsteps. He had joined the Napoleonic army in 1810 as a military surgeon and after arriving in Suriname he left the army and set up practice as a surgeon and obstetrician in Paramaribo. He was appointed city surgeon ('stadschirurgijn') in 1825 and became a member of the Collegium Medicum. However, his career in Suriname was cut short by his contact with leprosy sufferers. In 1828, he visited the new Batavia leprosy asylum and became so upset (he even claimed to have become partially paralysed) that he went on sick leave back to the Netherlands and never returned. His last years were spent as an obstetrician in the Dutch town of Lisse where he wrote about his upsetting experiences with leprosy.[32]

Van Hasselaar differentiated four phases in the development of leprosy. In the first phase, the symptoms are coloured spots, anaesthesia, weariness, and malodour. In the second phase, the anaesthetised spots turn into tubercles and nodes, accompanied by swelling and an increasingly bad body odour. In the third phase, extremities fall off, anaesthesia increases, and people become indifferent and incapable of labour. Plantation administrators requested that leprosy sufferers in this phase be sent to the leprosy asylum. In Van Hasselaar's opinion, this was too late and he held the administrators responsible for the contagion of half the slaves on their plantations because of the administrators' carelessness, miserliness, and wilfulness. For Van Hasselaar, they should have listened to the doctors and surgeons. In the fourth and final phase, the disease fully ravages the sufferer's body. Hands and feet fall off completely, cataracts blind the sufferers, and European sufferers' skin turns purple and blue.[33]

Van Hasselaar was full of humanitarian sympathy for sufferers of leprosy, but his attitude towards the Afro-Surinamese did not differ from Schilling or Kuhn. In his view, contracting the disease was far more difficult to bear for a white European than for the 'uncivilized' and 'unreasoning' black person, who was supposedly already happy if he could only exercise his animal lust.[34] Van Hasselaar's views of the slaves in Suriname decisively influenced his aetiology, and he feared that by spreading the disease, Africans threatened white dominance.

Van Hasselaar had no doubt that leprosy was contagious, but like Schilling he believed that the leprosy poison required a contributing factor. He suggested that the leprosy germs could be transmitted in many ways, even by touch or breath. However, the most important transmission mechanism to Van Hasselaar was sexual intercourse. According to Van Hasselaar, people with a dissolute lifestyle, who without second thought made use of opportunities to satisfy their 'animal lusts' were affected by leprosy sooner or later. Thus, Van Hasselaar further sexualized the discourse on leprosy. This explained why one could live close to a leprosy sufferer for a long time without contracting the disease, and why leprosy did not affect the majority of people living in Suriname; these people did not have sexual contact with leprosy sufferers.[35] For instance, the Roman Catholic bishop Grooff, who often spent time among the leprosy sufferers in Batavia, was never affected by the disease.[36] To Van Hasselaar, the implication was that the Bishop was not affected because he did not have sexual intercourse with black women in Batavia.

Van Hasselaar devoted much attention to what he perceived as the sexual aspects of leprosy. He thought that in the second phase, leprosy sufferers became increasingly sexually active. He was fascinated by this and expounded in some detail about how a sufferer could perform the sexual act an 'incredible' number of times in one night.[37] Van Hasselaar believed that even in the final and lethal fourth phase, animal lust did not diminish. Not only did the leprosy sufferers drink and eat well, but their sexual urges increased as well. Van Hasselaar suggested that suffers regularly performed the sexual act only one hour before death.[38] Because of their sexual vigour, many healthy women preferred a husband who was leprous above a healthy one, if this Dutch surgeon is to be believed. According to Van Hasselaar, at night the streets of Paramaribo were filled with leprous women attempting to seduce all men they encountered. He warned young European men to be careful and control their urges. Van Hasselaar gave a warning example of a young naval officer who was ensnared by a young and pleasant but leprous girl. The two had already had intercourse several times before Van Hasselaar 'did his duty' and warned the officer, who was much surprised and 'scared stiff' since the 'cunning voluptuary' (the girl) had succeeded in hiding her deformed hands under her shawl.[39]

The descriptions of black female leprosy sufferers in the writings of Van Hasselaar and other Europeans in Suriname contrast sharply with a rare description of a white female leprosy sufferer. Afro-Surinamese women were often portrayed as having loose morals, being lascivious, and even outright prostitutes. However, in an 1866 novel, the Protestant clergyman Cornelis van Schaick gave a radically opposing image of white Mathilde. She was a young white woman of a bourgeois family who was tragically diagnosed with leprosy, forced to break off her engagement to a captain, and isolated in the family home. Mathilde was pure and saint-like, unlike the Afro-Surinamese women depicted by Van Hasselaar.[40] While leprosy was not necessarily a disease of the poor or the lower classes or races in the nineteenth century, race, class, and status were decisive in stigma and the sort of intervention deemed necessary.

Van Hasselaar connected the fact that one could be a carrier of leprosy poison, but not develop the disease, to his ideas about Afro-Surinamese racial inferiority. He believed racial inferiority stimulated their fatalism and indifference to leprosy. To the Afro-Surinamese, God's will ('Gado nannie') decided who will be affected by the disease: 'Elfe Gado no gie mie, mie no sal kissie' ('If God does not give me the disease I will not get it'). According to Van Hasselaar, slaves were fatalistic, lazy, and slovenly with insufficient personal hygiene and a diet containing too much unhealthy salty fish. Last, but not least, slaves believed in a 'treef', a taboo animal that everyone receives at birth, which cannot be eaten or otherwise used without the direst consequences including contracting leprosy.[41] As Chapter 3 will show, this belief was of paramount importance to the way in which Afro-Surinamese managed health and disease. The consequences for Van Hasselaar were dire: 'the [belief in] predestination and its resulting carelessness to protect oneself against contagion, bring many to [contribute to] the expansion of the contagion'.[42] Van Hasselaar also considered another racial connection with leprosy: Jews were more at risk than other white people because their diet was similar to the Afro-Surinamese and included a great deal of unripe bananas and salty fish.[43] However, Van Hasselaar did not attribute the extreme negative racial qualities to the Jews that he attributed to the Afro-Surinamese.

For Van Hasselaar, the Afro-Surinamese kept the danger of contagion alive because of their fatalism and laziness. One result was the existence of a group of highly sexually charged Afro-Surinamese women. Their lust found a ready outlet because of the animal urges of European men. Since there was a scarcity of European women in the colony, men's lust was not compensated by the latter's 'civilizing gaze'.[44] European men in Suriname were in danger of losing their self-control and health, and thus ultimately colonial authority was at stake. Van Hasselaar essentially developed Schilling's position. He ended his book by warning about the danger of leprosy returning to Europe, and advocated for a medical examination of all Europeans returning to the Netherlands.[45] He even suggested a kind of proto-eugenics by prohibiting the marriages of black leprosy sufferers.[46]

Schilling and Van Hasselaar's framing of leprosy had an impact beyond the boundaries of Suriname. Their influence was noticeable in medical textbooks in the Netherlands, which educated Dutch doctors in the racial and sexual aspects of the disease. Van Hasselaar was a major source for the chapter on leprosy or 'Surinamese boasie' in Wilhelm Büchner's 1840 medical textbook. Büchner understood the following from writings about Suriname. Proven and established medical knowledge about leprosy was that in the final phase of the disease the animal lust of the sufferer increased, hunger became insatiable, and the sufferer often engaged in sexual intercourse a few hours before death. The disease could be inherited and contagious in hot climates. The main transmission mechanism was sexual. At risk were Europeans who lived in the tropics and could not control their sexual and other animal urges, and Africans and Jews because of their unhealthy diet. Jews were as much at risk as Africans. Since medicine could do little to stop or cure the disease, segregation (as in Suriname) was the appropriate policy. However, since Büchner lived in the Netherlands where leprosy sufferers were hard to find, he thought that leprosy was not contagious in the temperate European climate.[47]

The influence of Dutch treatises on leprosy was not confined to the Netherlands. Two decades later, one of the most influential medical scientists of the nineteenth century, the German Rudolf Virchow, referred to both Van Hasselaar and an 1841 medical dissertation on elephantiasis by J. P. ter Beek as well as other Dutch medical publications.[48] Not everybody, however, agreed with the opinions on leprosy from Suriname, even though these views were internationally influential. For example, the Göttingen professor Conrad Fuchs did not believe in a racial predisposition. In 1840, he observed that leprosy was actually a disease of the lower classes and an economic and social rather than a racial problem.[49] The Surinamese racial framing of leprosy was influential, but ultimately only of relevance in the colonies.

Treatments

The therapeutic optimism that Schilling expressed in 1769 had completely disappeared in Suriname half a century later. There was no cure for leprosy. Van Hasselaar described all failed methods of treatment, including sulphur baths and purgatives, decoctions of juniper or wild rosemary, and the administration of sarsaparilla or mercury.[50] These were typical remedies used in colonial medicine in the West Indies and tried in a variety of diseases. Only one of these remedies introduced by a British physician in 1758 seemed to promise results: oral administration of cinchona and sassafras together with the application of blistering agents on affected body parts. Van Hasselaar tried this remedy after returning to Europe. He treated a thirteen-year-old girl with discoloured skin patches, which he diagnosed as a symptom of leprosy. According to Van Hasselaar, the treatment worked.[51] However, this was only an incidental success story and the cure was never practised on a large scale.

Despair about finding a cure led Europeans to try pills made from the flesh of a green lizard. A widespread belief among the Afro-Surinamese held that eating or even touching a lizard could cause leprosy, and this was partly supported by the belief in the existence of the treef, the supernatural taboo animal. Europeans considered that this belief system was superstitious, but for pragmatic reasons they were prepared to entertain the notion that there was some connection between leprosy and lizards. 'Anything that to some extent could serve to cure or even to alleviate the disgusting 'walgelijke ontzettende kwaal' ('disgusting, terrible disease') is such a thing to be wished for that we will give it a place in our journal', was the opening of an article in the Paramaribo weekly *Nieuwe Surinaamsche Courant* ('*New Surinamese Journal*') in 1834. The article claimed that the cure for leprosy had already been known for decades among Europeans in the French Caribbean. A letter from a plantation manager on the island of St Kitts dated 1800 described how one of his slaves had cured herself of leprosy with pills made from green lizard flesh. Since then, the planter claimed that he had successfully administered the pills to all of his slaves with signs of leprosy.[52] The *Nieuwe Surinaamsche Courant* took this story at face value and Van Hasselaar also mentioned this cure in his book on leprosy, but thought it was all nonsense.[53]

Another cure that failed was homeopathic. In January 1828, the Surinamese government gave permission to the German physician Constantin Hering to set up a hospital for the cure of 'onreine' ('impure') diseases: skin diseases such as yaws and leprosy.[54] Hering was an adamant follower of Samuel Hahnemann, the founder of homeopathy, and he would later become a pioneer of the homeopathic movement in the United States. In 1826, the King of Saxony had sent Hering on a botanical expedition to Suriname. Here, Hering moved in the circles of the Protestant Moravian Church that had been sending missionaries to Suriname from their headquarters in Herrnhut, Germany since 1735. Hering also wrote about homeopathic medicine, leading to a falling-out with the King of Saxony. When setting up his own medical practice, Hering became convinced that he had found a homeopathic cure for leprosy. Since he had successfully treated the wife and daughter of the governor of Suriname for other ailments, Hering was in good standing with the colonial government, despite the negative views of other doctors and surgeons on homeopathy. In addition, because there were so few doctors in Suriname, Hering's services were welcome. In 1832, Hering was a member of a medical committee, consisting of three people, for an investigation into cholera, a new disease threatening the colony. On his request he was allowed to establish a clinic for patients with yaws and leprosy on the grounds of the Kwatta plantation. However, Hering left Suriname for the United States on New Year's Eve 1833 because of his wife's death.[55]

In Hering's view, leprosy was an expression of 'psora', one of the three 'chronic miasma' that homeopathy considers to be responsible for human diseases. Psora is a latent and dynamic factor of the human constitution and is expressed in skin diseases in particular. According to homeopathic principles, 'like cures like'. For leprosy to be cured, therefore, the patient would have to be infected with a similar disease, such as psoriasis, a chronic skin disease characterized by scaling patches of various sizes. Hence, Hering gave his patients 'antipsoric' substances based on or akin to sulphur. In accordance with the rule in homeopathy, the dosage was minimal. He treated twenty-four patients and claimed that although none was completely cured, five were considerably improved after treatment. However, from his descriptions it remains unclear whether these patients really suffered from a form of leprosy.[56]

Hering's writings on his treatments are suffused with the same racial and sexual connotations of leprosy as those of Van Hasselaar. Hering complained about the African slaves' fatalism and lifestyle. He even claimed that the slaves had a bad influence on their masters.[57] A plantation owner with a great interest in the natural sciences gave a twenty-year-old slave girl with leprosy named Pauline to Hering to experiment on. Hering first started to administer sulphur. After nine weeks when the patches had decreased in size, but the nodes had grown and the itching had become more severe, he switched to the administration of sepia. Though Hering later considered that the dosage had been a little too high, he claimed that it 'worked'. After eight weeks of further treatment, physicians examining Pauline did not diagnose her with leprosy. Hering's presentation of Pauline's case story shows how he framed leprosy within a homeopathic perspective, but with characteristic elements of colonial framing reminiscent of Van Hasselaar. He describes how Pauline became sexually active with both the European and the mulatto overseer of Hering's establishment immediately after her 'cure'. He noted that this pointed to the dangers of the sexual attraction between the leprous slave girl and a European overseer.[58]

Hering's experiments on Kwatta stopped when he left for the United States. However, the colonial government kept hoping for a cure. In the 1840s, they sponsored a number of experiments on slaves by Dutch physicians. From 1843 to 1846, C. G. N. Gravenhorst tried an unsuccessful therapy of a nutritious diet, steam baths, the administration of aloe as a purgative and potassium hydroxide for purification, and the external application of croton oil.[59] A similar experiment by his colleague F. Nolte was started in 1843, but discontinued in 1845 because of the lack of success.[60] In 1847 and 1848, experiments with guano (bat dung) again had negative results. Experiments with the Sandbox tree (*Hura crepitans*) (used by Brazilian natives as an emetic) on six military leprosy sufferers, soldiers, and non-commissioned officers by the chief military physician in Suriname, Schorrenberg, also failed.[61] Additional experiments over the next decade failed as well.[62] Since the search for a cure was unsuccessful, prevention became of even greater importance in colonial medicine and to the colonial government.

Prevention and segregation

From the moment the Dutch regained possession of Suriname in 1816, government officials, doctors and surgeons asked time and again for stricter policies for leprosy prevention and segregation. To begin with, the colonial government decided that the Voorzorg leprosy settlement was situated too close to the surrounding plantations. A new leprosy asylum named Batavia was founded in a far more isolated place on a riverbank near the mouth of the Coppename River and surrounded by jungle. In 1824, a new edict made examination and prevention of leprosy an explicit task for the Collegium Medicum, the medical supervisory board. Furthermore, changes in Dutch policies towards slaves made a new legal structure for leprosy control necessary. Before 1828, slaves were legally considered to be the objects and possessions of their owners, thus, slaves could be investigated and sent to an asylum without problems. In 1828, their legal position changed and they formally become persons, so new legislation was necessary. The new leprosy edict of 1830 functioned as the legal basis for leprosy policies for the next hundred years. It was built on the foundations of the edict of 1790 that had first formulated the principle of compulsory segregation. A new element of the 1830 edict was concern with not only leprosy, but also elephantiasis. Both were seen as contagious diseases that had to be controlled with the same methods. A Committee of Investigation ('Commissie van Onderzoek') for the diseases was installed. Its members were the Paramaribo city physician, city surgeon, and two physicians who were thought to have specialist knowledge of the diseases. A high-ranking government official without medical training presided. The Paramaribo police received extensive powers of criminal investigation to identify suspected leprosy and elephantiasis sufferers through a general visitation of all dwellings in Paramaribo once every three years, starting in 1831. Bounties were promised for every slave with leprosy that was brought before the Committee of Investigation. Teachers were ordered to check on school children.

The Committee of Investigation had to establish to which of the following three categories people belonged: (1) sufferers of leprosy or elephantiasis; (2) non-sufferers; or (3) people suspected of being sufferers, but without a conclusive diagnosis. Persons in the first category were transported to the Batavia asylum, except for the more wealthy Europeans (including Jews), who could remain segregated in their homes. The 'innocent' in the second category were released. If the suspects were slaves or poor whites, they were brought to Boniface, a special government terrain near Paramaribo. After one year, they were examined again to determine whether to send them to Batavia or to return the slaves to their owners and set the poor whites free.[63] The edict provided a hierarchy of confinement: the slaves or the poor were sent to the jungle asylum, but not the better-off whites.[64] Leprosy policy remained as much a matter of labour management as public health.

The edict of 1830 put leprosy control in Suriname on a firm basis. But how did it manage the messy realities of everyday health and disease control? Some claimed that it was a failure. The Dutch physician Charles Landré (1806–1892) published figures in 1858 on the numbers of investigations for the years 1831 to 1855. Landré worked as a physician in Suriname from 1840 to 1861, and was a key player in the execution of leprosy control policies. In 1846, he was first mentioned as a member of the Committee of Investigation.[65] In 1856, he became the city physician for Paramaribo.[66] His son Charles Louis Drognat Landré (1844–1917) published the data for a few more years (until 1859) and included it in his 1867 MD thesis on leprosy in Suriname. The father and his son Landré used these data to support their contention that leprosy control in Suriname had been a failure and that the edict of 1830 had never been properly executed. However, a close examination of their data reveals how the pattern turns out to be more complex. (See Table 1 and Figure 1.)

There are peaks in the years 1831–1832, 1839–1843, and 1846–1847. The first peak is unsurprising, since the edict of 1830 had just been put in place and there was much catching up to do. The reasons for the second peak are not clear. The third peak might have been connected to the stimulus of Landré's membership of the Committee of Investigation and the discovery of new cases of leprosy among the military. In 1844, two Dutch officers had to be repatriated because they were affected with leprosy.[67] The year 1831 is the only one apart from 1847 in which the number of investigations of suspects rose above 200.

The Committee spent quite some time on investigations that numbered (on average) 100 per year. Examinations took place in the state hospital of Paramaribo. Officially, all four medical members of the Committee had to make a separate investigation of the suspect behind closed doors, and then report their diagnosis in writing to the secretary without consulting the other members. However, in Landré's experience (from 1846 onwards) the doctors did not bother with a formal written report.[68] When agreement about the diagnosis was not forthcoming, a verbal discussion took place among the doctors and ultimately a majority vote decided.[69] If the suspect was declared infected and was a slave, they could not leave the hospital, but were prepared for transportation to Batavia. A free person had to go back to their dwelling

Table 1 Investigations of the Committee of Investigation, 1831–1859

Year	Number of investigations of suspects by Committee	Number of infected persons
1831	320	194
1832	193	53
1833	100	35
1834	73	32
1835	78	31
1836	103	41
1837	85	48
1838	85	52
1839	192	66
1840	189	63
1841	121	53
1842	123	51
1843	172	59
1844	82	27
1845	81	48
1846	126	57
1847	234	150
1848	68	29
1849	100	49
1850	44	25
1851	45	24
1852	63	37
1853	53	24
1854	45	18
1855	47	23
1856	31	12
1857	79	26
1858	31	6
1859	32	11
Total	2995	1344

Source: Landré, 'Naschrift', p. 233; Drognat Landré, *Besmettelijkheid*, p. 34.

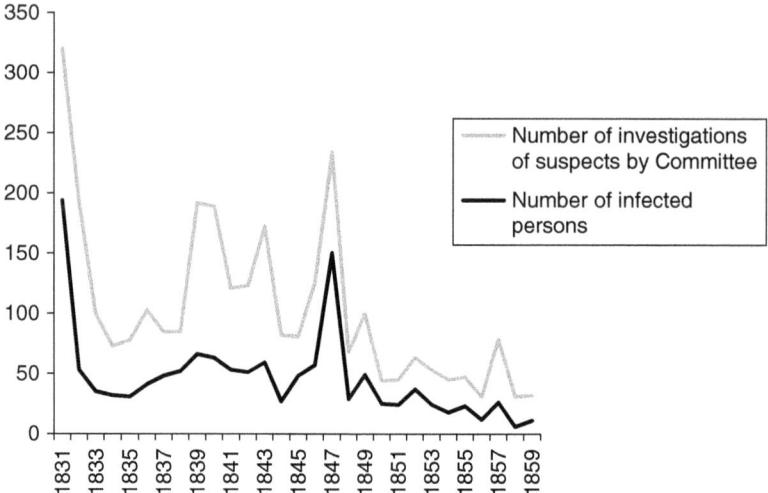

Figure 1 Investigations of the Committee of Investigation, 1831–1859

and remain there in isolation. If the Committee remained uncertain of the diagnosis and a slave was declared suspect, he or she would have to carry a suspect sign around their neck.[70] In all, 1,344 persons were declared to be infected from 1831 to 1859, which was close to 45 per cent of all examined persons. Only in the peak years of 1831 and 1847, were the number of 'convictions' higher than seventy, which exceeded 50 per cent of all investigations.

Since there were an average of 100 investigations per year, then on average in any given year, one in fifty to fifty-five of all people in Suriname would have been brought before the Committee. Most of the suspects were slaves, ensuring that the average among the slave population was even higher. However, these figures are only averages over a long period. For any given year, the situation varied, as it did for any plantation or district. For instance, on the Somersorg plantation in 1830, five of the eighty-four slaves (6 per cent) were sent to Batavia.[71] A survey of the records of eleven plantations turned up seven plantations from which slaves had been sent to Batavia between 1830 and 1863. Only one of the eleven plantations had records for the whole period, and on this plantation there were no infected slaves at all. For all examined plantations the records show that a total of twelve slaves were sent to Batavia, one, two, or three at a time.[72] The impact of the 'Great Confinement' policy becomes clearer when comparing the total of 1,344 people diagnosed as infected over a period of thirty years with population figures. Before 1849, the figures are scant and not reliable. According to one count, the population in 1837 numbered 58,962 persons: 50,467 slaves and 8,495 freemen.[73] This number is probably too high, but even so, a relatively high number of these persons must have been examined since the introduction of the 1830 edict. In the period between 1831 and 1837, the Committee held 952 examinations, and 434 persons were declared infected. The total number of examinations was higher since there were examinations in addition to those by the Committee. Before being put up to auction, slaves had to be examined by the city physician. In the early 1830s, the search for the infected was so intense that it even endangered the slave market. Already in June 1831, a new edict softened the 1830 edit for the slave owners, because the 1830 measures could lead to 'discomfort among good residents'. Suspected slaves could once again be bought and sold, and when whole plantations were sold, the slaves did not need to be examined. Slaves sent to Batavia were no longer property of their owner. This meant that if the slaves ever returned from Batavia, the previous owner had no obligation to take care of them.[74]

Despite this easing of the 1830 edit, the search for infected persons remained rather intense, and not only among slaves. For instance, in 1833 the attorney general sent a list of thirty suspects to the Committee: twenty-eight of them were free people. Two streets in Paramaribo were described as focal points of contagion.[75] Military officers sent reports of suspected leprosy in the army ranks to the Governor General, including the widow and children of a sergeant major in 1834, and even a captain in 1835. The Governor General then ordered further investigations by the Committee.[76] There is no evidence that the initial search for the infected was as lax as father and son Landré (who were not present at the time) would later suggest.

It is not known how many of the 'infected' were ultimately transported to Batavia in the 1830s and 1840s. Van Hasselaar mentioned that approximately 200 people were living in Batavia when he visited in 1828. Other accounts of the 1820s speak of approximately 300. In 1833, a head count in the asylum numbered 401 infected sufferers, of whom 397 were slaves and four freemen.[77] Since almost 250 people had been declared infected by the Committee in the two preceding years, the population of Batavia had probably doubled since Van Hasselaar's visit.

There are no data about whether people sent to Batavia were suffering from leprosy or elephantiasis until the second half of the 1850s.[78] In this period, the number of elephantiasis sufferers among those infected was always at least half of the number of leprosy sufferers. There were no slaves among elephantiasis sufferers despite claims in 1841 by the Dutch military surgeon J. Hille that the streets of Paramaribo were full of elephantiasis sufferers: his estimate was 2,000 out of 20,000 inhabitants. He also claimed that as long as the signs of elephantiasis or leprosy were 'local' – for example, affecting the legs or the scrotum – they were not of much interest to the Committee doctors. However, when the doctors observed open infection or disfigurement that was not concealed by clothing, the diagnosis was leprosy.[79] In 1851, of the 424 sufferers living in Batavia who had been examined, only thirty-four were diagnosed with elephantiasis (one with both elephantiasis and leprosy).[80] Leprosy, a visible disease among slaves, and not elephantiasis, was the primary target of the Committee of Investigation.

There was more criticism of the Committee of Investigation's examinations. Hering thought that the Committee members did not wish to spend much time with the suspects.[81] Landré claimed that often people with psoriasis, a skin disease that is not life threatening, were mistaken for possible leprosy sufferers and declared suspect although they were acquitted after re-examination.[82] In 1832, the auctioneer of the slave auction in Paramaribo complained about the behaviour of then city physician E. P. Schlörholtz. The auctioneer sent six slaves to the doctor's house to be examined before the auction. After he had done so three times, the doctor had still not examined the slaves. When the doctor finally came around to the examinations, he performed them in a nonchalant way. Finally, he sent a female slave Magdalentje and her child François to the Committee of Investigation although according to the auctioneer they showed no signs of leprosy at all. The president of the Committee to whom the auctioneer complained in writing did not accept critiques by laymen and wrote in the margin of the letter: 'this does not need an answer'. The doctor Schlörholtz was, after all, a member of the Committee.[83]

In 1851, an examination of the leprosy sufferers in Batavia showed that of 304 people who had been transported there in the decade before, twenty-six (8.5 per cent) did not have either leprosy or elephantiasis. They were all slaves.[84] Slaves' social status was too low to ensure a thorough examination by European doctors and surgeons, resulting in the segregation of dozens of slaves who did not suffer from leprosy at all.

Advising stricter policies

In the 1840s, colonial authorities maintained the impetus of the 'Great Confinement' policy. In 1845, the regulations of the leprosy edict were tightened. Sufferers were no longer permitted to own patents or companies. Children had to show a health certificate before being allowed to enter school. Infected freemen were not allowed to trade, work, or hold public functions. The police were ordered to pick up sufferers when they were in the streets and to send them to Batavia.[85] However, the effects of the segregation polices remained unsatisfactory. In January 1847, the Committee for Medical Investigation and Supervision or CGOT ('Commissie voor Geneeskundig Onderzoek en Toevoorzigt'), the new medical regulatory board that had taken the place of the Collegium Medicum in 1838, reported on leprosy to the Governor General of the Dutch West Indies. The CGOT estimated the number of leprosy and elephantiasis sufferers in Suriname to be approximately 2,500. This meant that almost 5 per cent of the population was infected.[86] A total of 464 sufferers were then living in Batavia. If the CGOT estimates were correct, only one in five carriers of the disease were segregated.

The CGOT further claimed that almost everyone in Suriname was indifferent to the situation of leprosy. The report stated the following: Afro-Surinamese were fatalist, more wealthy citizens were afraid of segregation, the great majority of the population was unmarried and lived in sin and promiscuity (so increasing the risk of contagion), and the plantation owners were negligent and sometimes even gave money to leprosy sufferers to set up a shop instead of turning them in to the Committee of Investigation. For the CGOT, visitation policies were not executed correctly and after dark the streets of Paramaribo filled with leprosy sufferers. There were sufferers repatriating to the Netherlands without medical examination. Inspection of schoolchildren was incomplete since the doctors only examined the children's faces. The Batavia asylum itself was a source of contagion. Segregation was totally insufficient since people left the asylum at will to gather food and even went to Paramaribo for weeks. The CGOT also accused the police of not cooperating. Before 1836, the police had not taken an active part in the detection of sufferers, so the CGOT claimed, and it was the slave owners and parents who brought suspects before the Committee. Although this had changed, the CGOT claimed that over the previous fifteen years, the police had brought in only six sufferers. What the CGOT did not mention was that the police might have had other priorities. The police force of Paramaribo was quite small and expanded from only six agents in 1800 to just thirty in 1862 on the eve of the abolition of slavery.[87]

Faced with the perceived laxness of almost everyone apart from the CGOT, the committee rung the alarm bell and recommended even stricter measures, which included more rigorous observance of the regulations, a public information campaign by the police, and yearly visitations to all dwellings and schools. Additional recommendations included founding a second Batavia for female sufferers to separate the sexes and arrest sexual reproduction by those with a hereditary disposition for leprosy, and increasing propaganda for marriage so that people would not feel the need for extramarital sexual relations with potential leprosy sufferers. Finally, a medical examination of all people returning to the Netherlands and in the Netherlands of all people coming from Suriname was suggested.[88]

The Governor, Baron van Raders, became quite worried because of this advice and its alarming tone. He again issued an edict that demanded health certification for schoolchildren.[89] Not knowing what more to do, he sent the CGOT recommendations to the Minister of Colonial Affairs in the Netherlands, who in turn sent it to the Royal Institute of the Sciences. In December 1847, a committee from the Institute, consisting of two professors of medicine (C. Pruys van der Hoeven from Leiden, and G. Vrolik from Amsterdam) and one former governor of Suriname (J. C. Rijk) made recommendations. Since the three had little knowledge of leprosy, they consulted a 1841 Latin MD thesis on 'Surinamese elephantiasis' by the Amsterdam physician J. P. ter Beek. Ter Beek had never been to Suriname, but summarized the published literature on the topic, and foremost Schilling's work. To Ter Beek, climate, diet, lifestyle, and hereditary predisposition could all play a role in so weakening the constitution that the disease was expressed.⁹⁰ In this manner of thinking, Africans were particularly predisposed to leprosy and elephantiasis according to the committee of three, but Europeans were not, except when having sexual intercourse with Africans. In the European climate, leprosy was not considered to be contagious. Nonetheless, even the committee of three held that it was advisable to demand a health certificate for Europeans repatriating to the Netherlands. Other committee recommendations were explicitly based on the Bible, thus showing the committee's lack of expertise on the subject. They advocated better diet and personal hygiene, purification, and segregation of sufferers. However, for the committee it was essential to segregate European sufferers from other sufferers in Suriname, and separate sexes by building a second Batavia.⁹¹

Although the committee of three was more concerned about the situation in Suriname than the danger to the Netherlands, it still saw leprosy as a potential colonial danger. Seventy-five years after the publication of Schilling's MD thesis, his framing of leprosy as an African disease with sexual and moral connotations was still maintained by the Dutch medical community. There was a general consensus among the medical professors of the Dutch universities and the Surinamese medical community, as was shown when the committee of three's advice was discussed again by the CGOT and the Committee of Investigation in Paramaribo in June 1848. There was only disagreement about the role of diet. Landré emphasized the role of diet as a 'predisposing cause' of leprosy, but did not agree with the other members' idea that the consumption of pork was especially harmful. The latter, however, thought it wise to ban pig cultivation in Batavia, possibly because of the notion that many pigs suffered from leprosy, which prompted the prohibition of pig farming on government plantations the following year.⁹²

The next step in the procedure brought all the advice to the members of the Colonial Council ('Koloniale Raad'), an advisory council consisting of prominent citizens of the colony. Not everyone on the council agreed with the estimates of the number of leprosy sufferers. The attorney general ('procureur-generaal') De Kanter thought the number was only a thousand. Another member was less specific and claimed that in the forty years that he had spent in Suriname, he had seen plantations with no evidence of contagion, plantations where a third of the slaves were infected, and even a few plantations where half had become infected. The conflation between leprosy and elephantiasis did not make the situation clearer. Opinion was equally divided about the advisability of appointing a medical doctor to Batavia. According to De Kanter, since leprosy was incurable, what was the point of a doctor? He did support the idea of a segregated neighbourhood for leprosy sufferers to replace Batavia. But who was going to pay for this?[93]

In the Netherlands, the committee of three had the answer. The Governor sent all Surinamese reactions to the Minister of Colonial Affairs, who sent them on to the Royal Institute in January 1851. This time only two of the committee of three, Pruys van der Hoeven and Rijk, discussed the problem. Their advice was to let the sufferers pay for their new neighbourhood. The poor ones could do so by performing labour, the wealthy ones with money. Building the neighbourhood did not have to be costly since one could build dwellings of wood and not stone, and requisitioned slaves from the plantations could dig a canal that would separate the neighbourhood from the town.[94]

By 1850, fears about leprosy's contagion in Suriname were expressed in the Colonial Reports ('Koloniale Verslagen'). Since the introduction of a new and more liberal constitution in 1848, the government was accountable to Parliament, and the Minister of Colonial Affairs had to send in a yearly report on the situation in the colonies. In 1849, the first report on the situation stated that despite the strict policies, leprosy and elephantiasis had increased in Suriname, especially among the poorer classes.[95] Until 1857, this statement was repeated annually in the Colonial Report, together with remarks about the indifference of the poor to the danger of contagion.[96] The Colonial Report of 1857 further expressed worry about the existence of leprosy among the Maroons, who might bring contagion to the population that was under Dutch control.[97]

Despite all the advice and discussions, in the end no action was taken apart from the appointment of a medical doctor for Batavia. First, there was no agreement about the exact number of leprosy sufferers. Second, there was no available money. Therefore, despite all the alarm bells, no stricter or more thorough policies were put in place from 1851 to 1861. On the contrary, during this period, there was a decline of the 'Great Confinement' policy connected to a decisive change in the economy – the approaching end of slavery.

The decline of the 'Great Confinement' policy

In 1849, the Committee of Investigation had a last burst of activity followed by two small peaks in 1852 and 1857. The Batavian population had dwindled from its probable all-time high in 1849. This happened not because the policy was revoked or modified, since even in 1853 and 1855 the leprosy edict had been reinforced. Now there were annual school investigations, obligations to report a sufferer within twenty-four hours, general police visitations of all dwellings were held as often as needed at the discretion of the commissioner of police, and a certificate of health was mandated for all requests for slave manumission.[98] These measures were mere paper tigers and the last gasp of the 'Great Confinement' policy in the age of slavery, because in reality the search for leprosy sufferers had declined.

This decline occurred to Landré's great disgust. In 1850, he withdrew from the Committee of Investigation and CGOT where he had been a full member since 1848.[99] His disgust was intensified by personal tragedy. One of his children became infected with leprosy, possibly as a result of contact with other children. In 1852, his son Drognat began to show signs of the disease at the age of two and died ten years later shortly before his father returned to Europe.[100] Thus, Landré became a vigorous critic of the laxity of Surinamese leprosy control. Although the figures of the 1850s seem to prove Landré correct, the figures of the 1830s and 1840s show another pattern. The priority of the fight against leprosy only decreased from 1855 onwards. By this year, leprosy sufferers were demanding re-examinations by the Committee of Investigation. For instance, in 1856, a fifty-year-old man who had been segregated for five years because he had shown signs of leprosy in his youth and was missing several fingers, demanded his release. Closer investigation showed that the man's condition had not deteriorated since his segregation. His case was seen as a matter of 'spontaneous cure' and he was released in 1858.[101]

How can this change in the execution of policies be explained? Had the fears of leprosy been exaggerated, and were most leprosy sufferers actually found and diagnosed by the end of the 1850s? Or were other factors at play? One might suppose that the shift was related to rising medical doubts about the contagious nature of leprosy. This may have been triggered by the medical study *Om Spedalskhed* ('On Leprosy') by the Norwegians Boeck and Danielssen that was published in 1848 in a French translation and made accessible to the international medical community. The Norwegians denied the role of contagion in the aetiology of leprosy and gave a primary role to heredity. Historians of leprosy have come to the conclusion that within twenty years, Danielssen and Boeck's publication led to a paradigm shift in medical thought about leprosy that culminated in an influential 1867 report by the British Royal College of Physicians. Instead of placing emphasis on the contagious nature of the disease, or even on the flexible and 'plastic' combinations of contagion, hereditary predisposition, lifestyle and climatological factors, this report constructed leprosy as a hereditary disease and denied the role of contagion.[102]

Did this change in medical views have an influence on medical discussions in Suriname and the Netherlands in the age of slavery? The German L. L. A. Deutschbein was appointed as the first medical doctor in Batavia at the end of 1850 and his report on the asylum is important in answering this question. Deutschbein had received his MD thesis at the University of Halle in 1840.[103] Having a spirit of adventure, five years later he was a medical practitioner in Suriname; however, this was not as a medical doctor, but rather as a surgeon in the Upper Cottica district in the interior.[104] Something must have gone wrong with his career in Europe. Perhaps he took the position in Batavia in an attempt to promote his career. Unfortunately, he died the following year, in July or August 1851.[105] Just before his death, Deutschbein finished writing a report on the situation of the leprosy sufferers in Batavia.

In this report, Deutschbein devoted special attention to the discussion of the role of heredity. However, his ideas show no influence of the ideas of Danielssen and Boeck. This was not because Deutschbein was unaware of the medical literature. For instance, he used the differentiation between 'lepra mutilans' (anaesthetic leprosy characterized by spots or maculae and potentially lethal), 'lepra tuberculosis' (tubercular leprosy characterized by nodes and tubercles and less dangerous than lepra mutilans), and elephantiasis or roos as used in European medical textbooks.[106] Deutschbein asked whether it was true that leprosy skips a generation and is transmitted to grandchildren. In 1851 there were three healthy African women living in Batavia who had come from Voorzorg. They had children and grandchildren in Batavia; some were infected while others were not. Of a total of 121 people born in Batavia, only twenty-six had leprosy, which was not even one in five. But though heredity seemed to play a minor role, leprosy did not seem especially contagious judging from Deutschbein's observations. He found healthy men or women living with infected partners and having healthy children with them.[107]

Deutschbein's successor in Batavia was Ooijkaas, a former military doctor and city physician of the town of Utrecht. Ooijkaas arrived in Batavia in 1853. He confessed that before coming to Batavia he knew next to nothing about either leprosy or elephantiasis. However, this did not stop him from performing another detailed examination of all sufferers. As Deutschbein had done, Ooijkaas too looked for hereditary factors and came to similar conclusions. Of 315 infected persons, thirty-three (10.5 per cent) had leprosy in the family. Therefore, heredity was of minor importance. Ooijkaas explicitly quoted Schilling in his analysis and wrote that climatological factors and the negligence of the colonial government in the fight against the contagion had made leprosy a significant problem in Suriname.[108]

In Suriname, there were no changes in the medical perspective on leprosy in the 1850s and early 1860s. Nor was leprosy's contagiousness, at least in the Surinamese climate, doubted in the Netherlands. An 1857 MD thesis by Karel Schönfeld at the University of Groningen (basically a review of the literature on leprosy) stated that the disease was hereditary, but within a complex aetiology in which other factors could play a role. For Schönfeld, the disease was especially present among the lower classes. The disease was contagious in many parts of the world but not in Europe where it had lost this property.[109] Only one publication denied the contagiousness of leprosy in Suriname as well. The naval medical officer S. Friedman, who had visited Suriname, published a book in 1861 in which he expressed doubts about the contagiousness of the disease. He had never seen a European sailor infected with leprosy.[110] However, Dutch doctors and surgeons in Suriname did not share Friedman's opinion.

The decline in the 'Great Confinement' policies cannot be explained by changes in medical opinion. There is one other factor that might explain the decline. From its beginning in the 1790s, the 'Great Confinement' had been aimed primarily at slaves, and was not only a medical policy, but also a kind of strategy for labour management. It was exactly in this area that profound changes to the slave economy of

Suriname occurred in the 1850s. In 1853, a committee was installed in the Netherlands to prepare for the abolition of slavery and the emancipation of the slaves, though it took another decade (until 1863) to actually be realized. Did the disappearance of the slave economy run parallel with an increasing lack of interest in the public health dangers supposedly posed by the slaves? Or was there a purely monetary consideration that was of importance? Plantation owners and other slave owners were vital in bringing suspected slaves before the Committee of Investigation. Emancipation would give them financial compensation for their slaves, but not for all of them. In 1857, an early proposal of a Dutch government committee for the Emancipation Act specifically excluded slaves with leprosy or elephantiasis from owner compensation.[111] In the final version of the act, a committee of three physicians appointed by the governor was to examine all slaves suspected of having leprosy or elephantiasis. If one year after the examination these slaves had not been declared healthy, there would be no financial compensation.[112]

A major reason for the decline in the 'Great Confinement' policy since 1855 could have been that slave owners were attentive to the outcome of the discussions surrounding financial compensation. As long as the outcome was undecided, slave owners feared losing possible upcoming financial compensation for any slaves they sent away. Once again, it would have been in their economic interest to hide the signs of leprosy in their slaves and not to cooperate with the policies of the colonial state.

Conclusion

After the Emancipation Act of 1863, all 33,922 freed slaves in Suriname were medically examined. By that time, medical examinations in Batavia had already revealed that a significant number of the sufferers in Batavia did not have leprosy or even elephantiasis at all; many of these people had been born there. In Batavia, 273 freed slaves were diagnosed with leprosy. Of the 33,560 freed slaves living outside Batavia, 107 were also diagnosed with leprosy. This meant that 1.1 per cent of the total former slave population was afflicted with the disease. If one assumed that this percentage was approximately the same for the 21,987 who were freed persons in Suriname before the Emancipation Act (many of them mulattos and freed slaves), this would mean that there were an additional 244 leprosy sufferers in Suriname, and a total of 624 sufferers in the whole population, of which only 273 had ended up in Batavia.[113]

From a present-day medical perspective in the twenty-first century, these figures might point to leprosy's limited risk of infection. To observers in 1863, the figures indicated that the messy realities and interests of everyday life obstructed the proper execution of public health policies. However, this judgement overlooked the fact that despite this messiness, in the period between 1830 and 1855, the detection and segregation of leprosy sufferers in Batavia was quite intense based on an eighteenth-century medical heritage of racializing and sexualizing the disease, and based on the fears of leprosy's threat to white dominance and colonial order. The result was a 'Great Confinement' policy that was primarily directed at slaves and the poor, which was as much about labour management as public health policy. Ending slavery meant ending the need for this kind of labour management. Although the legal structure of the 'Great Confinement' was not changed, its execution disappeared with the emancipation of the slaves.

Notes

1 M. Foucault, *Madness and Civilization: A History of Insanity in the Age of Reason* (London: Routledge, 2001), pp. 38–40.
2 J. F. E. Einaar, *Bijdrage tot de kennis van het Engelsch tussenbestuur van Suriname 1804–1816* (Leiden: M. Dubbeldeman, 1934).
3 F. A. Kuhn, 'Over de elephantiasis te Suriname', *Hippocrates* 7 (1828), pp. 12–28; A. van Hasselaar, *Beschrijving der in de kolonie Suriname voorkomende elephantiasis and lepra (melaatschheid)* (Amsterdam: S. de Greber, 1835), p. 17; C. H. Landré, 'Naschrift bij P. Duchassaing, Over de Elephantiasis Arabum in West-Indië', *West-Indië. Bijdragen tot de bevordering van de kennis der Nederlandsch West-Indische kolonien*, 2 (1858), pp. 222–33.
4 Landré, 'Naschrift', p. 226.
5 B. R. Laurence, '"Barbadoes leg": Filariasis in Barbados, 1625–1900', *Medical History* 33 (1989), pp. 480–8.
6 Kuhn, 'Over de elephantiasis', p. 12; Van Hasselaar, *Beschrijving*, p. 17.

7 Elephantiasis is now known as filariasis and clearly distinguished from leprosy. Its cause, a parasite, was discovered in the 1880s.
8 A. P. Kuijs and G. J. Rijnders, *Waarneming eener Elephantiasis aan het linkerbeen* (Amsterdam: Ten Brink & De Vries, 1820).
9 C. Hering, 'Vorläufige Mittheilungen über die auf Surinam einheimische Lepra', in *Herings Medizinische Schriften*, vol. 1 (Göttingen: Ulruch Burgdorf, 1988), pp. 11–21, on p. 16.
10 Kuhn and Rijnders, *Waarneming*, p. 21; J. P. ter Beek, Dissertatio medico-inauguralis de elephantiasi Surinamensi (MD thesis, University of Leiden, 1841).
11 Kuhn, 'Over de elephantiasis'.
12 C. L. Drognat Landré, *De besmettelijkheid der lepra arabum, bewezen door de geschiedenis dezer ziekte in Suriname* (Utrecht: J. L. Beijers, 1867), pp. 22–3.
13 L. C. van Panhuys, 'De Gouverneur-Generaal Willem Benjamin van Panhuys', *De West-Indische Gids* 6 (1925), pp. 291–320, on pp. 301–2.
14 F. Oudschans Dentz, *De kolonisatie van de Portugeesch Joodsche natie in Suriname en de geschiedenis van de Joden Savanne* (Amsterdam: S. Emmering, 1975), p. 20.
15 Drognat Landré, 'Besmettelijkheid der lepra', p. 19. De Mattos is a Portuguese Jewish name: Oudschans Dentz, *Kolonisatie*, p. 10. On racism against Jews in the Caribbean: P. Pluchon, P., *Nègres et juifs au XVIIIe siècle. Le Racism au siècle des Lumières* (Paris: Tallandier, 1984).
16 H. Buddingh', *De geschiedenis van Suriname* (Amsterdam: Nieuw Amsterdam, 2012), p. 117.
17 Hallewas, 'Gezondheidszorg', pp. 60–1.
18 The most detailed account of the Surinamese health care system from this perspective is G.-J. Hallewas, G.-J., 'De gezondheidszorg in Suriname' (PhD thesis, Groningen University, 1981).
19 Surinamese newspapers (hereafter SN): Advertisement Collegium Medicum 9 November 1832, *Surinaamsche Courant*, 11 November 1832.
20 A. Kleinman, *Patients and Healers in the Context of Culture: An Exploration of the Borderland between Anthropology, Medicine, and Psychiatry* (Berkeley, CA: University of California Press, 1980).
21 Description of the Surinamese healthcare system: F. A. Kuhn, *Beschouwing van den toestand der Surinaamsche plantagieslaven. Eene oeconomisch-geneeskundige bijdrage tot verbetering deszelven* (Amsterdam: C.G. Sulpke, 1828).
22 Surinamese almanacs (hereafter SA): *Surinaamsche Almanak voor het jaar 1820* (Paramaribo and Amsterdam: Beijer/Sulpke, 1819).

23 In the preface to Kuhn, *Beschouwing van den toestand der Surinaamsche plantagieslaven* Kuhn speaks of his eleven years' experience in Suriname.
24 Kuhn, *Beschouwing*, pp. 38–9.
25 SN: Advertisement G. Bergman in *Surinaamsche Courant* 17 July 1828. Kuhn is no longer mentioned in the *Surinaamsche Almanak voor het jaar 1829* (Paramaribo: Maatschappij Tot Nut van 't Algemeen, 1828) (SA). He died the same year in the Netherlands (*Surinaamsche Courant*, 23 September 1828) (SN).
26 Kuhn, *Beschouwing*, pp. 53–4.
27 Kuhn, *Beschouwing*, p. 41.
28 Kuhn, *Beschouwing*, pp. 40–2.
29 F. van Heshuysen, 'Memoire sur la forme du gouvernement de Surinam et de la nature de chaque employ', in Einaar, *Bijdrage*, p. 195.
30 Kuhn, *Beschouwing*, pp. 36–8.
31 Kuhn, 'Over de elephantiasis'.
32 Van Hasselaar, *Beschrijving*, pp. 1, 23, 63; SA: *Surinaamsche Almanak voor het jaar 1820*; *Surinaamsche Almanak voor het jaar 1825* (Paramaribo: Maatschappij Tot Nut van 't Algemeen, 1824); *Surinaamsche Almanak voor het jaar 1830* (Paramaribo: Maatschappij Tot Nut van 't Algemeen, 1829); *Surinaamsche Almanak voor het jaar 1828* (Paramaribo: Maatschappij Tot Nut van 't Algemeen, 1827); *Surinaamsche Almanak voor het jaar 1831* (Paramaribo: Maatschappij Tot Nut van 't Algemeen, 1832); *Surinaamsche Almanak voor het jaar 1832* (Paramaribo: Maatschappij Tot Nut van 't Algemeen, 1833); SN: *Surinaamsche Courant* 17 May 1825, 23 December 1830, 24 December 1833; *Opregte Haarlemsche Courant* 20 March 1838.
33 Van Hasselaar, *Beschrijving*, pp. 27–38.
34 Van Hasselaar, *Beschrijving*, p. 3.
35 For example, Van Heshuysen, 'Memoire', in Einaar, *Bijdrage*, p. 195.
36 Van Hasselaar, *Beschrijving*, pp. 5, 38–9.
37 Van Hasselaar, *Beschrijving*, p. 34.
38 Van Hasselaar, *Beschrijving*, p. 38.
39 Van Hasselaar, *Beschrijving*, pp. 29–34.
40 C. van Schaick, *De manja, Familietafereel uit het Surinaamsche volksleven* (Arnhem: D. A. Thieme, 1866).
41 Van Hasselaar, *Beschrijving*, pp. 3–4, 10, 38–41.
42 Van Hasselaar, *Beschrijving*, p. 10.
43 Van Hasselaar, *Beschrijving*, pp. 26–7.

44 A. L. Stoler, *Carnal Knowledge and Imperial Power: Race and the Intimate in Colonial Rule* (Berkeley, CA: University of California Press. 2002).
45 Van Hasselaar, *Beschrijving*, pp. 73–6.
46 Van Hasselaar, *Beschrijving*, pp. 23–5.
47 W. F. Büchner, *Geneeskundig handboek voor beginnende kunstoefenaren* (Amsterdam: H.J. Berntrop, 1839), pp. 336–45.
48 Though he found their disease descriptions confusing: R. Virchow, *Die krankhaften Geschwülste. Dreissig Vorlesungen*, I (Berlin: August Hirschwald), pp. 296–9.
49 C. H. Fuchs, *Die krankhaften Veränderungen der Haut und ihrer Anhänge* (Göttingen: Dietrichsen Buchhandlung, 1840), pp. 641–3.
50 Van Hasselaar, *Beschrijving*, pp. 68–72.
51 Van Hasselaar, *Beschrijving*, pp. 68–72.
52 SN: *Nieuwe Surinaamsche Courant en Letterkundig Dagblad*, 19 August 1834.
53 Van Hasselaar, *Beschrijving*, pp. 68–72.
54 Drognat Landré, 'Besmettelijkheid der lepra', pp. 28–9.
55 F. Oudschans Dentz, 'De. Constantin Hering en Christiaan Johannes Hering', *West-Indische Gids* 12 (1931), pp. 147–60; K.-H. Gypser, 'Constantin Hering – Versuch einer Biographie', in Gypser (ed.), *Herings Medizinische Schriften*, pp. xi–xliv; *Surinaamsche Courant* 3 Septetmber 1832; 29 December 1832.
56 C. Hering, 'Vorläufige Mittheilungen', in Gypser (ed.), *Herings Medizinische Schriften*, vol. 1, pp. 11–21; C. Hering, 'Fernere Mittheilungen über die Lepra und ihre homöopathische Heilung', in Gypser (ed.), *Herings Medizinische Schriften*, vol. 1, pp. 22–47; C. Hering, 'Die antipsorische Mittel in ihrer Beziehung zur Lepra', in Gypser (ed.), *Herings Medizinische Schriften*, vol. 1, pp. 113–219; C. Hering, 'Einige Bemerkungen über das Psorin', in Gypser (ed.), *Herings Medizinische Schriften*, vol. 1, pp. 388–422.
57 Hering, 'Fernere Mittheilungen', in Gypser (ed.), *Herings Medizinische Schriften*, vol. 1, pp. 24–5.
58 Hering, 'Fernere Mittheilungen', Gypser (ed.), *Herings Medizinische Schriften*, vol. 1, pp. 26–38.
59 'Mededeelingen nopens de *lepra* in onze West-Indische bezittingen', *Nieuw Praktisch Tijdschrift voor de Geneeskunde* 28, Nieuwe reeks 1 (1849), pp. 546–68, 761–70, on pp. 546–8, 567.
60 'Mededeelingen nopens de *lepra*', pp. 546–53.

61 *Verslagen over de lepra te Suriname* (Amsterdam: G. M. P. Landonck, 1851); J. Karbaat, 'Sociaal-geneeskundige beschouwingen over de personeelsleden van de troepenmacht in Suriname en hun gezinnen' (MD thesis, Leiden University, 1963), p. 189.
62 SN: *Surinaamsche Courant*, 1 May 1855; Colonial Reports (hereafter CR) 1855, p. 28; SN: *De Curacaosche Courant* 26 May 1859; 'Berigten', *West-Indië. Bijdragen tot de bevordering van de kennis der Nederlandsch West-Indische koloniën*, 2 (1858), pp. 296–7.
63 *Gouvernementsblad* [hereafter GB] nr. 13, 7 September 1830.
64 For a similar situation in British India: Buckingham, *Leprosy in Colonial South India*, pp. 157–88.
65 SA: *Surinaamsche Almanak voor het jaar 1847* (Paramaribo: Maatschappij tot Nut van 't Algemeen, 1846).
66 CR 1856, p. 24.
67 Karbaat, 'Sociaal-geneeskundige beschouwingen', p. 189.
68 Landré, 'Naschrift', p. 229.
69 National Archive, The Hague, Algemene secretarie Nederlandsche West-Indische bezittingen in Suriname 1830–1847 [hereafter AS], Nr. 10, Minutes Committee of Investigation 22 August 1835.
70 J. Hille, 'Ueber die *Elephantiasis*; nach eigenen Beobachtungen in West-Indien', *Wochenschrift für die gesammte Heilkunde* (1841), pp. 433–42, on p. 438.
71 A. A. van Stipriaan Luïscius, 'Surinaams contrast. Roofbouw en overleven in een Caraïbische plantage-economie' (PhD thesis, Vrije University Amsterdam, 1991), p. 388.
72 E. Klinkers, 'De bannelingen van Batavia. Lepra-bestrijding gedurende de negentiende eeuw in koloniaal Suriname', *OSO* 22 (2003), pp. 50–61, on p. 50.
73 SA: *Surinaamsche Almanak voor het jaar 1838* (Paramaribo: Maatschappij tot Nut van 't Algemeen, 1837), p. 141.
74 GB 14 June 1831, nr. 12.
75 AS 10, Attorney General to the government secretary and chairman of the Committee of Investigation, 13 March 1833.
76 AS 10, letters 1 February 1834 and 17 October 1835.
77 AS 12.
78 Drognat Landré, *'Besmettelijkheid der lepra'*, p. 34.
79 J. Hille, 'Ueber die *Elephantiasis*', pp. 433–42, 457–63.

80 L. L. A. Deutschbein, 'Report', *Tijdschrift voor de Wis- en Natuurkundige Wetenschappen* 5 (1852), pp. 100–5.
81 Letter C. Hering to A. Sapf, Suriname, 28 September 1827, in Gypser (ed.), *Herings Medizinische Schriften*, pp. 1–10, on pp. 9–10.
82 C. Landré, 'Bijdragen tot de kennis der ziekten van de negers in de kolonie Suriname', *Nieuw Praktisch Tijdschrift voor Geneeskunde in al haren omvang*, 31 (Nieuwe reeks 4, 1852), pp. 496–7.
83 AS 9, Letter vendumeester Keyser to the Committee of Investigation, 12 November 1832; SA: *Surinaamsche Almanak voor het jaar 1834* (Paramaribo: Maatschappij tot Nut van 't Algemeen, 1833).
84 Deutschbein, 'Report'.
85 GB 1845, nr. 3, nr. 12.
86 CR 1849.
87 E. Klinkers, *De geschiedenis van de politie in Suriname, 1863–1975. Van koloniale tot nationale ordehandhaving* (Amsterdam: Boom, 2010), p. 17.
88 Report CGOT 28 January 1847, in 'Mededeelingen nopens de *lepra*', pp. 554–65.
89 GB 1847, nr. 7.
90 Ter Beek, 'Dissertatio', pp. 8–9, 20–2.
91 Report of the committee Pruys van der Hoeven, Vrolik and Rijk in *Verslagen over de lepra te Suriname*, pp. 2–15.
92 SN: *Curacaosche Courant*, 17 November 1849.
93 *Verslagen over de lepra*, pp. 19–21.
94 *Verslagen over de lepra*, pp. 22–30.
95 CR 1849, p. 216.
96 CR 1850, p. 147; 1851, p. 202; 1852, p. 195; 1853, p. 256; 1854, p. 302; 1855, p. 28; 1856, p. 24; 1857, p. 24.
97 CR 1857, p. 495.
98 GB 1853 nr. 3; 1855 nr. 8.
99 SN: *Surinaamsche Courant en Gouvernments Advertentie Blad* 17 June 1848, 5 January 1850.
100 T. May, 'De lepra, haar voorkomen, verspreiding en bestrijding, in 't bijzonder in Suriname', I, *West-Indische Gids* 8 (1927), pp. 547–56, on p. 550; H. Menke, S. Snelders and T. Pieters, 'Omgang met lepra in 'de West' in de negentiende eeuw. Tegendraadse maar betekenisvolle geluiden vanuit Suriname', *Studium* 2 (2009), pp. 65–77, on p. 75. The case is described by the boy's brother, Charles Louis Drognat Landré, without mentioning the family connection: Drognat Landré, *Besmettelijkheid der lepra*, pp. 53–5.

101 Drognat Landré, *Besmettelijkheid der lepra*, p. 31.
102 R. Edmond, *Leprosy and Empire: A Medical and Cultural History* (Cambridge: Cambridge University Press, 2006), pp. 44–60; M. Vollset, 'Globalizing Leprosy: A Transnational History of Production and Circulation of Medical Knowledge 1850–1930' (PhD thesis, University of Oslo, 2013).
103 L. L. A. Deutschbein, 'De noma infantum' (MD thesis, Halle, 1840).
104 SA: *Surinaamsche Almanak voor 1846* (Paramaribo: Maatschappij tot Nut van 't Algemeen, 1845); *Surinaamsche Almanak voor 1847* (Paramaribo: Maatschappij tot Nut van 't Algemeen, 1846).
105 SN: *Surinaamsche Courant en Gouvernements Advertentie Blad*, 21 August 1851.
106 Fuchs, *Die krankhaften Veränderungen der Haut*, pp. 634–93, 702–13; C. Canstatt, *Handbuch der medicinischen Klinik*, 3rd rev. ed., vol. 2 (Erlangen: Ferdinand Enke, 1855), pp. 1–15.
107 Deutschbein, 'Report'. A summary was published in *Nederlandsch Weekblad voor Geneeskundigen* 2 (1852), pp. 95–6.
108 V. Schneevoogt, 'Verslag op het rapport van den heer Ooijkaas, omtrent het lepreuzen etablissement Batavia, in de kolonie Suriname', *Verslagen en Mededeelingen der Koninklijke Akademie van Wetenschappen* 2 (1854), pp. 381–8.
109 K. D. Schönfeld. 'Verhandeling over de lepra in 't algemeen, en de elephantiasis tuberculosa in 't bijzonder' (MD thesis, University Groningen, 1857).
110 S. Friedman, *Nederlandsch Oost- en West-Indië, volgens de nieuwste inrigting, met betrekking tot aardrijkskunde, statistieken, voortbrengselen, luchtgesteldheid, en vooral tot den gezondheidstoestand* (Amsterdam: J.C.A. Sülpke, 1861), p. 222.
111 This discussion was reported on in the newspapers in the Dutch Caribbean, for example, SN: *De Curaçaosche Courant*, 22 August 1857.
112 *Handelingen Staten-Generaal*: www.statengeneraaldigitaal.nl/ (hereafter SG), Kamerstuk 1861–1862, nr. XXXV, ondernummer 18.
113 Drognat Landré, *Besmettelijkheid der lepra*, pp. 37–8.

3

Slaves and medicine: black perspectives

To Dutch doctors and surgeons, the beliefs of the African slaves and especially their belief in the treef as a taboo animal, motivated the slaves' laziness and fatalism and hindered the proper realization of medical policies and treatments. Their opinion was a one-sided and prejudiced view that ignored existing Afro-Surinamese health practices. In 1769, when Schilling wrote about his observations of leprosy treatment among the slaves in Suriname, he claimed that in general there was no treatment, and that the slaves worked until the signs of the disease became so severe their masters left them to their fate. A leprosy sufferer could not expect much from white medical professionals. At best, the sufferers would be segregated in their home, or at worst sent to Voorzorg or Batavia. However, from a black point of view there was another, spiritual aetiology and explanation for leprosy and other diseases, and proprietary cures and treatment methods. Schilling did point to the existence of an alternative option. Slaves turned to what he described as 'Ethiopian quacks', by which he meant African healers in general. He suggested that the African medical practitioners, freedmen or slaves, were in possession of secret herbal cures. These healers were found on all plantations.[1] There was an 'internal' health system practised and accessed by Afro-Surinamese, just as there was an internal economy wherein the slaves operated relatively independently of their masters.[2] This internal sector partially overlapped with white colonial medicine on a level of social interaction, such as plantation medical care or treatment of the same patients. There was also knowledge and information exchange between and within medical cultures when European surgeons and doctors tried to establish the value of non-European cures, or when Africans shared knowledge of cures from

Africa as well as America among themselves and with Amerindians. For instance, the slaves in the New World recognized plants with medicinal properties that they had used in Africa and passed this information on to each other.[3]

Afro-Surinamese leprosy management was part of a broad spectrum of healthcare activities, just as it was for the Europeans. The American psychiatrist and medical anthropologist Arthur Kleinman has modelled this spectrum in his healthcare system analysis.[4] Kleinman suggests that there is a 'special cultural system' rooted in arrangements of social institutions and patterns of interpersonal interaction. For Kleinman, a healthcare system cannot be understood as simply the development and progress of diagnosis and treatment dictated by biomedical sciences. The system is composed of a multitude of cultural notions, such as disease origins and the names that determine treatment choices and evaluations, as well as social structures and power relations. Kleinman's concept of a healthcare system integrates all 'health-related components' and views both patients and healers as embedded in these cultural meanings and social relations. His perspective of a healthcare system does not position a 'professional' sector of medical practitioners trained in biomedical sciences as opposed to a 'folk' sector of alternative healthcare. Rather, these sectors overlap with each other and with the 'popular' sector of patients, families, and communities.

When using Kleinman's terminology to analyse multiple historical periods and societies, the terms 'professional' and 'folk' sector need to be treated with some degree of scepticism. Many folk healers are as 'professional' as their 'professional' counterparts. Yet, Afro-Surinamese healthcare activities were to some extent declared illegal by the colonial rulers, especially their religious and magical activities, which Europeans feared. Nonetheless, in Suriname, as elsewhere in the Caribbean, enslaved Africans became agents of their own medical care and retained, adapted, and developed their own cultural notions in what historian Katherine Bankole has called a 'pursuit of holistic healing'.[5] Furthermore, in slave medicine, health activities became a means for Africans to gain a degree of agency in their lives.[6]

This chapter explores leprosy management on the black side of the medical market in the age of slavery as contrasted with the white side explored in the previous chapter. Unfortunately, any exploration of the black side of the medical market is limited and impressionistic since period sources generally offer perspectives filtered through the eyes of European observers. The slaves were not so much passive in their health behaviour as distrustful of their masters' medicine, and were not inclined to share their knowledge with Europeans. In searching for strategies to break through this silence, the historian runs the risk of uncritically using data from anthropological research and folklore studies, as if folk medicine is a static and unchanging field without historical nuances and dynamics. However, this is not to say that an attempt to understand folk medicine is necessarily doomed. Some historians have shown that an alternative perspective can be taken with promising results by using extant colonial sources and reading from a more 'bottom-up' perspective.[7] Valuable insights can be obtained when exploring sources on leprosy management in the age of slavery from this approach.

The natural: herbal medicine

How did the eighteenth-century female African herbalist mentioned by Schilling manage to cure more than twenty boasie sufferers? Schilling was a man disdainful of African medical practices, but he was also an adventurer-scientist with a thirst for useful knowledge, and therefore was prepared to buy the woman's secret cure. However, he wrote down his observations within the framework of his own cultural notions: as methods of purging, sweating, and applying ointments, the kind of treatments that a European practitioner understood. The secret cure started with the administration of a purgative twice a week. Schilling identified it as 'Gitte Gom'. This term shows how European and African practices overlapped, since what the Dutchman knew as Gittegom (Garcinia gumma-gutta L.) was not native to the Americas, but rather an Indian tree used medicinally in Asia as a purgative whose use was publicized in Europe in the sixteenth century and appeared in pharmaceutical handbooks.[8] When the female healer, whose name remains unknown, judged the patient to be sufficiently purged she gave the patient a hot concoction of the bark and roots of the Tondin tree three times a day. The Tondin was a tree that grew in swampy areas and was unknown to European botanists. Dutch ethno-botanical researcher Tinde van Andel identified Tondin from Schilling's drawing of the tree as Paullinia pinnata, a liana that is found in South America as well as

West Africa. In Suriname, the liana is now named 'feififinga' ('five fingers').[9] The patient drank the concoction, performed physical exercise and was then covered in blankets to sweat. After a few weeks of this regimen, the healer would apply a thick black ointment prepared from Viltkruid with lemon juice to the afflicted spots on the body. Viltkruid or English dodder (Cuscuta Americana, or another species of Cuscuta) is a parasitic plant that was also used in Surinamese Winti. Winti is the name of the African slaves' traditional religious and medical belief systems and practices that incorporate Native American practices and which were exchanged among slaves and Maroons through contact on the plantations and elsewhere. Dodder was used to suck the life out of its host.[10] If the spots disappeared and the last scabs fell off, the healer would declare the patient cured. This healing process lasted three to four months. Schilling had seen many leprosy sufferers cured of their spots, although he was told that in some patients with an unhealthy lifestyle the symptoms returned.[11]

This healer clearly was a dress mama – a female black healer using medicine and magic (the boundary line was diffuse) and with a high status and reputation among the Africans. In 1787, plantation manager Anthony Blom wrote about the male and female 'lukumen' (seers and magicians) who called themselves doctors. According to Blom, the slaves held lukumen sacred. At the sick bed, the lukumen 'pretended' to be able to predict whether the patient could be cured with all kinds of 'jests' (in Blom's words), such as looking into the sun. However, lukumen did not only prophesise, they also administered herbal medicine.[12] Schilling wrote that the healers experimented with all kinds of herbs, tasted them, and slowly discovered both the beneficial and adverse effects by administration and self-administration.[13] In the early 1820s, Adriaan Lammens, a Dutch judge in Suriname, sketched a similar picture of black medicine. Lammens, like Blom, regarded the prophecies and 'secret ceremonies' that the lukuman used as superstition and confidence tricks. Although Lammens did acknowledge that the slaves had some knowledge of herbal medicine, he considered it unwise to trust too much in this knowledge.[14]

Despite the scepticism of some Dutch observers, Europeans did consult black healers. The Jewish savants of Paramaribo were as derogatory of the African 'quacks' as Schilling was; however, they regretfully mentioned that a black 'priestess' named Dasina and other Africans were often consulted as prophetesses and healers by Europeans, especially European women.[15] Leading Europeans consulted the famous and notorious lukuman Quassie, who reputedly had cured himself of boasie.[16] One European leprosy sufferer sent a plant that he believed cured leprosy to the Governor of Suriname. In an accompanying letter, the man told this story. In 1810, as an administrator, he had entered into an affair with a mulatto girl of seventeen or eighteen years of age. Both he and the girl developed anaesthetic spots or maculae on their body. In Paramaribo, European doctors could not help, but a black man had cured them both with an herbal concoction. Since then, the European had successfully treated several slaves with similar spots with this medicine, or so he claimed.[17] Native American cures also circulated to other ethnic groups in Suriname. For instance, at the end of the nineteenth century, the Catholic missionary C. van Coll met mulattos and Europeans with leprosy among the tribes in the jungle. They had come to medicine men in search of a cure.[18] Schilling's interest in the cures of other cultures was certainly not unique among Europeans.

The reasons behind the practice of Schilling's anonymous African female healer are not so clear. Were they similar to Schilling's own therapeutic principles, and did the cure perhaps incorporate European medicine? If so, this would have made her therapy a typical instance of Afro-Caribbean medical syncretism, by her borrowing remedies, knowledge, and techniques from another cultural group.[19] Or, did the African female healer only work with native herbs to a degree unknown to the Dutchman? Africans used their individual and collective memories to look for and to find American remedies that were similar or identical to the African remedies that had become unavailable to them after their passage into slavery.[20] Did the female healer also put a little bit of magic in her ointments? Feififinga could be both a medical and a magical remedy. Kuhn found that the slaves used magical explanations especially for rare diseases or those with long progression, but in general, they looked for the cause of diseases in a polluted body and corrupted body fluids like the Europeans did. In the same way as their white counterparts, black healers administered medicine to purge and sweat their patients, and both groups prescribed healthier diets.[21]

What the Africans really knew never became explicit or clear to the Europeans, however. The slaves distrusted their masters and were not eager to share their knowledge with their oppressors. In 1762, Philippe

Fermin complained that botanical information could not to be obtained from the slaves with money or with gentle words.[22] Almost seventy years later, Hering wrote that the Africans often had great empirical knowledge, for instance of possible leprous spots on children's bodies, but held it secret. The European physician could never trust what Africans told them.[23]

Ethno-botanical research has identified herbal cures for leprosy that still have a place in Surinamese folk medicine, such as the use of a concoction of the bark of the small kasyuma tree (Annona reticulate).[24] Chocolate tea made with the dried seeds of cacao (Theobroma cacao) was administered at the first signs of leprosy.[25] Amoraman, the herb Lycopodiella cemua, is reported as a remedy for leprosy, although its folk name derives from its use in love magic.[26] Boasiman is one of the folk names for the small tree, Zygia latifolia. 'Boasiman' literally means leprosy sufferer, and this might point to its use in cases of leprosy. Another name is Kokobebisonki.[27] The difference between 'dry' and 'wet' boasie was apparent to the Africans, and the dry variant was called 'kokobe'. Schilling gave the name 'kokobe' as a word for leprosy among the Coromantins, a general designation of Akan ethnic groups from Ghana such as the Ashanti.[28] By 1950, these words were still in use: boasie for what was then called lepromatous leprosy, and kokobe for the relatively benign tuberculoid leprosy.[29] Today, in the twenty-first century, Surinamese health authorities still make this distinction between kokobe (dry leprosy) and boasie (wet leprosy) in their health propaganda.[30]

The brothers Frederik (1876–1909) and Arthur Penard (1880–1932), who made extensive studies of Surinamese folklore, and who are reputed to have become leprosy sufferers themselves, reported that a bradibita drink was used against leprosy by the Afro-Surinamese at the end of the nineteenth century.[31] The manuscript of the Swedish adventurer and plantation owner Carl Gustav Dahlberg, compiled in approximately 1770, contained much information about the medicinal uses of Surinamese plants. Although Dahlberg does not mention leprosy, he does mention the use of the bradibita herb (Leonotis nepetifolia L.) against yaws and other skin diseases, so the herb might have been in use against leprosy as well in his time.[32]

How widespread the use of these herbal cures for leprosy was cannot be ascertained. Nor is it known when the use of herbal cures for the disease began. Were they used for all kinds of leprosy? Were they already used in the eighteenth century, or not until a (much) later date? Only contemporary reports, such as Schilling's in the late eighteenth century, can provide answers to these questions; but these reports are rare. What is certain is that African healers (whether slave or freed, male or female, born in Africa or in Suriname) experimented with a wide variety of herbs they recognized based on their experience in Africa. At the same time, African healers were not averse to incorporating European medicine such as Gittegom in their practices. This is far from unique and it is important not to misunderstand folk medicine as a static practice. On the contrary, an Afro-Caribbean medical system was and is a composite of dynamic practices, or as is written of Haitian Voodoo: 'a vital living body of ideas and behaviours carried in time by its practitioners and responsive of the changing character of social life'.[33]

The supernatural: the treef

In the Afro-Surinamese experience of the cosmos, man is not only a physical or biological being, but also a spiritual one. In Winti, the soul or kra (in the head), the dyoko (guardian demons or ancestors in the supernatural world), the yorka (the immortal spirit), and the takru sani (threatening evil demons) all have profound influence on an individual's health.[34] The anthropologist Gloria Wekker wrote: 'The self is vulnerable, porous, as it were, to dangers of many different kinds. For datra siki/doctor's illness, also called Gado siki/God's illness, one consults a doctor, as opposed to negre siki/negro illness, for which one has to consult a bonu [spiritual healer].'[35]

Of central importance in Afro-Surinamese belief systems is the concept of the treef (Surinamese: trefoe, trefu). Stedman became familiar with the word in the 1770s.[36] During his sojourn in Suriname, he had come in close contact with African slaves, both male and female, and received much medical information from them.[37] He also recorded what he designated as 'superstitions' among the Africans. One was 'a direct prohibition in every family, handed down from father to son, against the eating of some one kind of animal food, which they call treff; this may be either fowl, fish, or quadruped, but whatever it is, no negro will touch it'. He contrasted this strict behaviour to the laxity of 'some good Catholics [who] eat roast-beef in Lent, and a religious Jew devouring a slice from a fat flitch of bacon'.[38] In the 1850s, the plantation owner A. Coster noticed that his maid, a Maroon, would not only avoid the consumption of unscaled fish (her treef) but also refused to cook on a fire on which unscaled fish had been cooked. Another of his maids had a cow as her treef, and she did not eat meat prepared with cow butter.[39]

Stedman was not the first one to write about the belief in this taboo among African slaves. Dutch books on Suriname mentioned the belief as early as 1770.[40] Willem Bosman, a merchant of the slave-trading Dutch West India Company, wrote in the beginning of the eighteenth century in his account of the Gold and the Slave Coast of West Africa (the coastline of present-day Ghana down to Angola):

> Each of the Negroes has its own prohibited food, one doesn't eat sheep, another not goats, cows, pig, wild fowl, [animals with] white feathers, and so on; everyone has his own [prohibited food]. And this is not for a week, month, year, but during the whole life forbidden ... the son shall not eat what the father could not, and so similar the daughter not what the mother couldn't.[41]

Two hundred years later, a German expedition to the Bantu people at the Loango-Angola coast discovered that everyone had a taboo called 'tschina', mostly inherited from the father, but also acquired, imposed, or received from the mother's family. Violating the tschina could cause disease.[42] If the original West African word for taboo was tschina or kina, where did the word 'treef' come from? In the city of Amsterdam, a similar word was and still is used by Jewish inhabitants: tereifa. Tereifa designates all food that is not kosher or in accordance with the Jewish dietary laws, and therefore unsuitable for consumption. The simplest explanations for the origin of treef is that either the Europeans gave the word to the African practice of not eating taboo food, and it was taken over by the slaves, or that the slaves took the word from the Europeans to explain the practice in a way their masters could understand. A 1778 dictionary still gives the word for the taboo as tchinna, but not as treef; the first recorded use of the word treef in print is in 1783.[43]

The Dutch dermatologist and leprosy expert Robert Simons (1909–1966) who lived in Suriname from 1911 to 1927 and was educated at the medical school in Paramaribo suggested a more active Jewish involvement. He speculated that Jewish slave owners forced their slaves to observe Jewish dietary laws. The slaves understood the tereifa within the framework of their own taboos originating in West Africa.[44] T. May, another Dutch physician in Suriname in the 1920s, thought that a belief in the treef did not exist in the Coronie and Nickerie districts where Jews did not live.[45] In the 1920s, Dutch medical doctor and public health officer P. H. J. Lampe observed that in the Surinamese language leprosy was also know as the Jewish disease ('Yu-siekie'). Was this an inheritance from the eighteenth century? According to Lampe, many Afro-Surinamese not only believed that Jewish slave owners had prescribed their dietary laws to their slaves, but also that the Jews had actually brought leprosy to Suriname as well. Influential Jewish families were supposed to have obstructed the compulsory segregation of family members and so thwarted the fight against leprosy. Whereas Europeans had blamed the Africans for bringing leprosy to Suriname, Afro-Surinamese had another explanation: the Jews were the culprits. This idea might have been influenced by European anti-Jewish beliefs. Many Afro-Surinamese treef were tereifa food for Jews. Lampe suggested that the Jewish influence could explain the distinctiveness of the Surinamese belief in the treef, in its details unlike other taboo beliefs of African or Afro-Caribbean people.[46] In 1945, an Afro-Surinamese woman who came to Paramaribo to work as a domestic help was warned to be careful with the laundry of Jewish people who might have the Yu-siekie. Young women were advised not to mingle with Jewish families in which the disease might occur.[47]

To the Afro-Surinamese the origin of the treef is supernatural. The treef can be inherited or acquired. It can be an inheritance from one's father. The child learns about it from his or her mother, who knows the treef her husband has to observe. In addition, other treef can also be acquired. When a child is born, a treef can be revealed in a dream to a blood relative (mostly female) or to a family acquaintance. An old woman or a prophetess can arrive to inform the family of the treef of a newly born child. Other treef can be acquired at a later age: so when one acquires a god or gods, one also acquires the treef of these gods. The use of certain magical charms demands that one should observe certain treef. Continuing beliefs today in the twenty-first century are that Amerindian spirits living in Surinamese people, in particular in children, have to be controlled by the observance of certain treef.[48]

Hence, within this belief system, the real reason for the affliction with leprosy is supernatural; the sufferers have committed an act (violating the taboo) that displeases the gods. Melville and Frances Herskovits, two American anthropologists who did research in Suriname in the 1930s, noted that the 'penalty [for violating the taboo] takes the form of punishment with skin-disease; a mild form of eczema at first, which, if neglected and aggravated by continued disregard of the inherited trefu, develops into leprosy'.[49] The name kokobe means impure or taboo. But according to Simons, even speaking the words boasie or kokobe was to be avoided. Instead, one spoke of takru sibi, the ugly worm.[50] Just as with venereal disease, leprosy was considered a takru siki, or a bad disease.[51]

Since a disease may be caused by the violation of the treef, and since the treef has a supernatural origin, it is a lukuman and not a medical doctor who has to diagnose whether a violation has occurred.[52] The brothers Penard heard of one such consultation at the end of the nineteenth century:

> A woman had a leprous girl and went from one lukuman and obiahman [magical healer] to another to find a cure, or to know the trefu of the child. One of these lukuman, a woman, advised her to gather fifty head kerchiefs and to lay them at night at fifty different crossroads. The woman did what she was told. [But, the sceptical Dutchmen wrote,] as is self-evident, this did not cure the girl.[53]

In the 1820s, European doctors began to notice that in the Afro-Surinamese belief system, violations of the treef could lead specifically to leprosy. Kuhn discovered that when slaves developed discoloured spots or maculae on their bodies they attributed this to a violation of treef. They called these spots 'treefvlekken' (treef spots) and feared that they could develop into leprosy. From the viewpoint of twenty-first century medical knowledge, the appearance of these maculae does not signify much. They could have been the first symptoms of some kind of leprosy, or they could have been 'lotta spots' (as they are called in Suriname), a benign and non-contagious infection. Kuhn thought they were harmless.[54] Landré identified the treef spots as psoriasis, a skin disease that is not life threatening. According to him, there was a connection with gastric disturbances caused by the consumption of turtle, crab, or fish, which was treef for some sufferers.[55] On the contrary, Kuhn had never witnessed that consuming one's treef led to the development of maculae.[56] Van Hasselaar witnessed several experiments and came to the same conclusion. By the 1820s, Europeans were sufficiently interested in a belief in treef to perform experiments on their slaves by using them as human guinea pigs.[57] Marten Douwes Teenstra, who worked in Suriname as an inspector of public works from 1828 until 1833, wrote that the slaves were given their treef to eat by hiding it in other food. Contrary to Van Hasselaar, he claimed that these slaves broke out in treefvlekken (eczema?) and had convulsions.[58] As will be examined further in Chapter 8, the Dutch rejected all supernatural beliefs in the treef, but did leave the possibility open that diet and skin disease could have a causal relation. However, as Chapter 8 also shows, they never succeeded in shaking the foundations of a supernatural treef belief among the Afro-Surinamese.

Conclusion

The widespread belief in treef among the Afro-Surinamese combined with the slaves' unwillingness to disclose details of their medical and spiritual views and practices to their European masters contributed to European ideas about the slaves' presumed indifference and fatalism. However, given the nature of the medical market in Suriname in the age of slavery there was no reason why slaves would cooperate with their masters in executing policies of leprosy prevention and involuntary segregation. The slaves had their own belief systems about leprosy, including a spiritual perspective that radically differed from the European medical perspective. Slaves had their own healers and treatments for leprosy. The effectiveness of these practices is questionable, but so too were their masters' cures. Africans with leprosy could not expect much other than deportation to Batavia and separation of their families from the Europeans.

In the age of slavery, the difference in perspectives on leprosy between African slaves and European masters contributed both to the development of the 'Great Confinement' policies (with an aim to countermand the perceived African fatalism and indifference) and to the undermining of these policies, since the slaves saw nothing to gain by observing or adhering to the policies. By viewing leprosy politics in the age of slavery from a bottom-up perspective, the mismatch between the rulers' intentions and interventions and the needs and wishes of the ruled becomes visible.

Notes

1 G. W. Schilling, *Verhandeling over de melaatsheid* (Utrecht: J. C. ten Bosch, 1771), p. 74.
2 I. Berlin and P. D. Morgan (eds.), *The Slaves' Economy: Independent Production by Slaves in the Americas* (London: Routledge, 1995).
3 M. Laguerre, *Afro-Caribbean Folk Medicine* (South Hadley, MA: Begrin & Garvey, 1987); S. Snelders, *Vrijbuiters van de heelkunde. Op zoek naar medische kennis in de tropen 1600–1800* (Amsterdam: Atlas, 2012); T. van Andel, 'The reinvention of household medicine by enslaved Africans in Suriname', *Social History of Medicine* 28 (2015), doi: 10.1093/shm/hkv014.
4 A. Kleinman, *Patients and Healers in the Context of Culture: An Exploration of the Borderland between Anthropology, Medicine, and Psychiatry* (Berkeley, CA: University of California Press, 1980).
5 K. Bankole, *Slavery and Medicine: Enslavement and Medical Practices in Antebellum Louisiana* (New York: Garland, 1998), p. xi. On the process of developing new cultural systems by Afro-Americans: S. W. Mintz and R. Price, *The Birth of Afro-American Culture: An Anthropological Perspective* (Boston, MA: Beacon Press, 1992).
6 E. D. Genovese, *Roll, Jordan, Roll: The World the Slaves Made* (New York: Pantheon, 1974), p. 229.
7 K. K. Weaver, *Medical Revolutionaries: The Enslaved Healers of Eighteenth-Century Saint Domingue* (Urbana, IL: University of Illinois Press, 2006); P. F. Gómez, 'The circulation of bodily knowledge in the seventeenth-century black Spanish Caribbean', *Social History of Medicine* 26 (2013), pp. 383–402.
8 M. Houttuyn, *Handleiding tot de plant- en kruidkunde*, vol. 3 (Amsterdam: Lodewijk van Es, n.d.), pp. 6–7.
9 It is still used as folk medicine in both Ghana and Suriname. In Ghana the ground plant is used as a poultice for skin problems. In Suriname the plant has the reputation of having a powerful obiah (magical strength). Email from T. van Andel to the author, 31 January 2014. More details on feyfifinga in Suriname: T. van Andel and S. Ruysschaert, *Medicinale en religieuze planten van Suriname* (Amsterdam: KIT, 2011), pp. 438–9. The drawing of

Schilling was published at the end of his 'Animadversiones in Ouseelianam et additamenta ad suam de lepra dissertationem' in: J. D. Hahn (ed.), *De lepra commentationes* (Leiden: Abr. van Paddenburg, 1778), pp. 119–203.
10 Houttuyn, *Handleiding*, vol. 7, p. 379; Van Andel and Ruysschaert, *Medicinale en religieuze planten*, pp. 171–2. On plant medicine and magic in Suriname: T. van Andel and S. Ruysschaert, 'What makes a plant magical?: Symbolism and sacred herbs in Afro-Surinamese Winti rituals', in R. Voeks and J. Rashford (eds.), *African Ethnobotany in the Americas* (New York: Springer, 2011), pp. 247–84.
11 Schilling, *Verhandeling over de melaatsheid*, pp. 73–8. Latin names and descriptions of the plants: Hahn (ed.), *De lepra commentationes*, pp. 59–60 and after p. 203.
12 A. Blom, *Verhandeling over de landbouw in de colonie Suriname* (Amsterdam: J. W. Smit, 1787), p. 350.
13 Schilling, *Verhandeling over de melaatsheid*, p. 74.
14 A. F. Lammens, *Bijdragen tot de Kennis van de Kolonie Suriname. Tijdvak 1816–1822* (Amsterdam: Geografisch en Planologisch Instituut VU, 1982), p. 108.
15 *Geschiedenis der kolonie van Suriname … Door een Gezelschap van geleerde joodsche mannen aldaar*, vol. 2 (Amsterdam: Allart & Van der Plaats, 1791), pp. 53–67.
16 Snelders, *Vrijbuiters*, p. 188.
17 T. May, 'De lepra, haar voorkomen, verspreiding en bestrijding, in 't bijzonder in Suriname', II, *West-Indische Gids* 9 (1928), pp. 29–31.
18 C. van Coll, 'Gegevens over land en volk van Suriname', *Bijdragen tot de Taal-, Land- en Volkenkunde van Nederlandsch-Indië* 55 (1903), p. 464.
19 Laguerre, *Afro-Caribbean Folk Medicine*, pp. 22–6.
20 Laguerre, *Afro-Caribbean Folk Medicine*, pp. 16–20.
21 F. A. Kuhn, *Beschouwing van den toestand der Surinaamsche plantagieslaven. Eene oeconomisch-geneeskundige bijdrage tot verbetering deszelven* (Amsterdam: C.G. Sulpke, 1828), pp. 40–9, 63.
22 P. Fermin, *Nieuwe algemeene beschryving van de colonie van Suriname* (Harlingen: Volkert van der Plaats, 1770), pp. 192–3.
23 C. Hering, 'Antipsorische Mittel', in K.-H. Gypser (ed.), *Herings Medizinische Schriften* (Göttingen: Ulruch Burgdorf, 1988), vol. 1, p. 207.
24 Van Andel and Ruysschaert, *Medicinale en religieuze planten*, p. 39.
25 Van Andel and Ruysschaert, *Medicinale en religieuze planten*, p. 314.
26 Van Andel and Ruysschaert, *Medicinale en religieuze planten*, p. 493.
27 Van Andel and Ruysschaert, *Medicinale en religieuze planten*, pp. 260–1.

28 Schilling, 'Animadversiones', in Hahn (ed.), *De lepra commentationes*, p. 175. According to T. May, the word is of French origin; May, 'Lepra', vol. 1, p. 548.
29 R. D. G. P. Simons, *Lepra. De maligne contagieuze Morbus Hansen en de benigne niet-contagieuze Hanseniden* (Amsterdam: Van Holkema and Warendorf, 1948), p. 11.
30 www.health.gov.sr/media/120195/wat_is_lepra.pdf. Accessed 3 February 2014.
31 F. P. Penard and A. Penard, '"Surinaamsch bijgeloof". Iets over Winti en andere natuurbegrippen', *Bijdragen tot de Taal-, Land- en Volkenkunde van Nederlandsch-Indië* 67 (1913), pp. 157–89, on p. 183.
32 Snelders, *Vrijbuiters*, pp. 200–1; Van Andel and Ruysschaert, *Medicinale en religieuze planten*, p. 275.
33 Alfred Métraux, quoted in Weaver, *Medical Revolutionaries*, p. 110.
34 C. J. Wooding, 'Winti: Een Afroamerikaanse godsdienst in Suriname. Een cultureel-historische analyse van de religieuze verschijnselen in de Para' (PhD thesis, University of Amsterdam, 1972), pp. 122–53.
35 G. Wekker, *The Politics of Passion: Women's Sexual Culture in the Afro-Surinamese Diaspora* (New York: Columbia University Press, 2006), p. 98.
36 H. D. Benjamins, 'Treef en lepra in Suriname', *West-Indische Gids* 11 (1930), pp. 187–218, on p. 196.
37 Snelders, *Vrijbuiters*, pp. 174–8.
38 J. G. Stedman, J. G., *Narrative of a Five Years' Expedition against the Revolted Negroes of Surinam, in Guiana on the Wild Coast of South America from the Years 1772 to 1777* (Amherst, MA: University of Massachusetts Press, 1972), p. 365.
39 A. M. Coster, 'De boschnegers in de kolonie Suriname. Hun leven, zeden, en gewoonten', *Bijdragen tot de Taal-, Land- en Volkenkunde* 13 (1866), pp. 1–37, on p. 20.
40 Fermin, *Nieuwe algemeene beschryving*, p. 153; J. J. Hartsinck, *Beschrijving van Guiana of de Wilde Kust, in Zuid-Amerika*, 2 vols. (Amsterdam: Gerrit Tielenburg, 1770), pp. 992–3.
41 W. Bosman, *Nauwkeurige beschrijving van de Guinese Goud-, Tand- en Slavenkust* (Amsterdam: Isaac Stokmans, 1709), vol. 1, p. 144.
42 E. Peschuël-Loesche, *Volkskunde von Loango* (Stuttgart: Stecker and Schröder, 1907), pp. 455–65.
43 Benjamins, 'Treef en lepra', p. 190. J. van Donselaar, *Woordenboek van het Nederlands in Suriname van 1667 tot 1867* (Amsterdam: Meertens Insitituut, 2013), p. 224, gives the first recorded use of 'treef' in print as in 1783.

44 Simons, *Bijgeloof en lepra*, pp. 52–4. Biography of Simons: G. A. Lindeboom, *Dutch Medical Biography: A Biographical Dictionary of Dutch physicians and surgeons 1475–1975* (Amsterdam: Rodopi, 1984), pp. 1813–15.
45 May, 'Lepra', p. 550.
46 P. H. J. Lampe, 'Het Surinaamsche treefgeloof. Een volksgeloof betreffende het ontstaan van de melaatschheid', *West-Indische Gids* 10 (1929), pp. 545–68, on pp. 565–6.
47 Email from Mildred Caprino to the author, 15 January 2014.
48 H. D. Benjamins, 'Treef', in H. D. Benjamins and J. F. Snelleman (eds.), *Encyclopaedie van Nederlandsch West-Indië* ('s-Gravenhage: Martinus Nijhoff, 1914–1917), pp. 685–7; Lampe, 'Het Surinaamsche treefgeloof', pp. 563–4; M. J. and F. S. Herskovits, *Suriname Folk-lore* (New York: Columbia University Press, 1936), p. 36.
49 Herskovits and Herskovits, *Suriname Folk-lore*, p. 37.
50 Simons, *Bijgeloof en lepra*, p. 55.
51 Penard and Penard, 'Surinaamsch bijgeloof', p. 182.
52 Herskovits and Herskovits, *Suriname Folk-lore*, p. 59.
53 Penard and Penard, 'Surinaamsch bijgeloof', p. 160.
54 Kuhn, *Beschouwing*, pp. 36–7.
55 C. Landré, 'Bijdragen tot de kennis der ziekten van de negers in de kolonie Suriname', *Nieuw Praktisch Tijdschrift voor Geneeskunde in al haren omvang*, 31 (Nieuwe reeks 4, 1852), pp. 496–7, pp. 495–7.
56 Kuhn, *Beschouwing*, pp. 36–7.
57 A. van Hasselaar, *Beschrijving der in de kolonie Suriname voorkomende elephantiasis en lepra (melaatschheid)* (Amsterdam: S. de Greber, 1835), p. 30.
58 M. D. Teenstra, *De landbouw in de kolonie Suriname* (Groningen: Eekhoff, 1835), p. 200.

4
'Battleground in the jungle': the Batavia leprosy asylum in the age of slavery

Leprosy sufferers detected under the 'Great Confinement' policies, and particularly those who were slaves, were sent to the Batavia leprosy asylum. By segregating them from the outside world, the perceived threat of a potential spread of their infection to the slave society, higher social groups, and even the Netherlands, was controlled. The geographically isolated leprosy asylum in the Suriname colony performed an essential role in colonial society. The asylum also established a working relationship between the colonial state and the Roman Catholic Church. This was possible because of the ambivalence in governmental policies regarding the management of the Batavia asylum. The vast majority of the sufferers in the asylum were a special category of slaves; they were unproductive. Since the slaves only cost the government money, it was unwilling to invest resources in the asylum. For instance, although the asylum was established in 1824, medical services were only provided in the 1850s. To provide support and sustenance for the sufferers, a symbiotic alliance developed between the colonial government and a religious group on the margins of Surinamese society, the Catholics.

Much of the historiography of Batavia has concentrated on the role of the Roman Catholic Church, which took charge of the care of the sufferers. Writers have emphasized the Church's humanitarian and religious motivations, and extolled the dedication and self-sacrifice of the priests working and living in Batavia.[1] This praise is certainly justified, but there were other dimensions to life in the asylum. For the Catholics, Batavia was the ideal place to demonstrate that they could perform a useful and necessary role in colonial society. Their work in Batavia was of further importance in rallying support for their Surinamese mission among the Catholics in the Netherlands. In Batavia, the Catholics went a long way to instil a 'Christian leper identity' in the sufferers, who had to shed their former heathen cultural identity to receive the compassionate care of the Catholics. However, the situation in the Batavia asylum was a far cry from other leprosy asylums across the globe established by the early twentieth century. These other asylums maintained tight disciplinary regimes to regulate all aspects of the sufferers' lives and transform them into 'modern' citizens. In part, the more relaxed regime at the Batavia asylum was based on the fact that the Catholics did not manage the asylum.

At the same time, the Batavia asylum was a source of contestation. Limited financial support from the government meant that poor living conditions caused dissatisfactions among sufferers. Not all sufferers embraced a new Christian identity, and many cherished their everyday autonomy, leading to conflicts with the priests in the asylum. The Catholic presence was contested by another Christian group, the Protestant Moravians, who tried to compete with the Catholics in caring for the leprosy suffers in Suriname. Batavia was a battleground in the jungle where religious, political and social conflicts between Catholics and Protestants, Christians and non-Christians, took place.

A new alliance between the colonial state and the Roman Catholic Church

A contemporary described the November 1824 relocation of leprosy sufferers from the asylum in Voorzorg to Batavia in two boats:

> They went in the early evening, and outside of Paramaribo, they had to sleep at night in the open air, in rain and wind, around small camp fires, so these naked sick people tried to cover themselves with some leaves of the trees against the dampness and the cold; those who had not died watched dawn arrive benumbed and cold and were wrestling with misery and despair; now they had to crawl again in the boat, trembling with cold and in total want of the most necessary things of life (food and cover); while some, observing their deceased companions, prayed for their own death ... and others brought this moment forward by jumping into the river.[2]

Although the Voorzorg asylum had basically been a dumping ground for leprosy sufferers, the sufferers themselves had no wish to move out of their habitat. Humanitarian or medical reasoning did not motivate the move to Batavia. Instead, it was inspired by the colonial government's wish for a stricter execution of the 'Great Confinement' policy by more rigorously segregating the leprosy sufferers from contact with the outside world. According to the Catholic *Chronicles of Batavia*, a manuscript compiled during the course of the nineteenth century, Voorzorg asylum was insufficiently isolated from the nearby plantations and there were many complaints that so-called asylum 'malingerers' were defying regulations and restrictions and visiting those plantations. Therefore, the colonial government decided to relocate the leprosy village to an even more isolated place in the jungle. The Batavia asylum was established on a bank of the Coppename River and surrounded by jungle on the landside.[3]

Even today, it takes almost half a day to reach Batavia from Paramaribo by car and motorboat. In the 1820s, the journey took at least two days since there was no road. Visitors had to sail from Paramaribo to the mouth of the Coppename and then follow the river in a sloop.[4] To the government, Batavia seemed just the place for a new asylum. It was sufficiently accessible to send new leprosy and elephantiasis sufferers, and was thought to have few opportunities for escape.

The relocation did not go as smoothly as expected by colonial officials. Some asylum malingerers refused to move. In June 1823, the forty-three healthiest Afro-Surinamese sufferers, those still capable of work had been selected from the approximately three hundred Voorzorg residents and brought to Batavia to clear the riverbank and prepare for the relocation.[5] Their departure led to unrest among the sufferers left behind. When the time for relocation came, the colonial government set huts on fire in Voorzorg to force the sufferers to comply with the Batavia relocation. Yet, they still refused to move. Presumably, there was not a soldier or police officer who relished the thought of having to move a few hundred leprosy sufferers by force with the possible risk of contamination. At this point, Governor Abraham de Veer made a decision that would be crucial for the history of Batavia and leprosy care in Suriname in general. To ensure the compliance of the malingering sufferers, he forged an alliance with a religious group that stood outside the political and social establishment of Suriname – the Catholics.[6]

A few years earlier, in 1820, the Catholic priest and missionary Paulus Antonius Wennekers had asked the colonial government for permission to visit the asylum in Voorzorg and preach to the sufferers. This request was refused since the dominant Protestant establishment in both the Netherlands and its colonies did not respect Catholics.[7] Catholics were a marginalized group in Suriname, and had only received freedom of worship in 1816. The Catholics' discriminated position made them see opportunities for conversion among other discriminated groups that no one cared about, in this case the leprosy sufferers. In 1817, the Catholic Church established a mission to Suriname and sent Wennekers and his fellow priest Ludovicus van der Horst from the Netherlands to the colony. Confronted with the Voorzorg asylum's malingerer non-compliance, Governor De Veer changed his predecessor's policy. The Catholics were given permission to send a priest to Voorzorg and in return the priest was asked to convince the sufferers to cooperate with the relocation to Batavia.[8]

How De Veer and the Catholics could be so sure that the malingerers were open to persuasion by a priest is unclear. There may have already been a substantial number of Catholics in Voorzorg. Wennekers succumbed to the tropical climate and died in April 1823, so it was Van der Horst who went to Voorzorg in November of that year. He preached, baptized, and according to the *Chronicles of Batavia* persuaded the malingerers to give up their resistance to the relocation with his 'fatherly speech'.[9] In the meantime, another Catholic priest, Joannes Willemsen, had joined the mission. He visited Batavia between January and April 1824 to assess the situation, and then went on to Voorzorg to reassure the leprosy sufferers that everything would be all right if they complied with the transfer to Batavia. By then, about half of the approximately three hundred Voorzorg inhabitants were reported to be Catholic.[10] Thus, Catholic influence was sizeable and unopposed by other organized religious affiliations. The leprosy sufferers appeared to have had more faith in the words of Catholic priests than in those of government officials.

Leprosy asylums were the first places outside of Paramaribo where Catholics could preach among the slaves, since this was not allowed on the plantations. The leprosy sufferers' spiritual welfare was not all that was at stake. Conversion and gaining followers was an essential aim in strengthening the Roman Catholic Church's position in Surinamese society. A second essential aim was for the Roman Catholic Church to demonstrate to the colonial state and establishment that despite centuries of distrust they were loyal citizens of the Dutch colonial empire, and had a significant contribution to make. In this way, Catholic emancipation, which was still unresolved in the Netherlands, could be furthered in colonial Suriname.

On their side, the colonial state could use the Catholics to keep control of the slave population since it was believed that the slaves who followed priests were less likely to cause trouble. The relocation of the leprosy sufferers from Voorzorg to Batavia was the first test to see if the Catholics could be helpful and they were successful. The leprosy asylum became a means of building what anthropologist Karin Bijker has designated as a symbiosis between the Roman Catholic Church and the colonial state.[11]

Christianising Batavia

Martinus van der Weijden, a Catholic priest who had only recently arrived from the Netherlands, visited Batavia after the relocation from Voorzorg in September 1826. According to the *Chronicles of Batavia* the situation in Batavia was appalling. Three hundred sufferers were said to live in one shed, plagued by floods because the dam could not check the river.[12] Was the situation really so extreme or was this an exaggeration of Catholic propaganda? It is unlikely that there was 'one shed' at the time since the sufferers generally lived in small huts. The *Chronicles* emphasized that construction work was still incomplete, which questions Willemsen's earlier assurances to the Voorzorg malingerers. However, Van der Weijden's first priority was spiritual since the care for the leprosy sufferers went hand in hand with conversion. In this effort, he had the support of the government-appointed director of the settlement J. H. Hek, who was also a Catholic. During Van der Weijden's visit, a hut in the village was consecrated as a Catholic place of prayer and 120 sufferers were baptized. Apart from preaching the new faith, Van der Weijden took drastic action against non-Christian beliefs. A large tree standing on the riverbank was considered sacred in Afro-Surinamese religious practices. The priest had the tree felled and put a large wooden cross in its place, thus symbolizing the triumph of the Christian faith over the heathen gods. In another version of this story, the tree was felled during a storm and the priest used the opportunity to erect the cross.[13]

The felling of this kankantri (Ceiba pentandra L.) demonstrates the Catholics' perception of their task among the leprosy sufferers. The missionaries believed that they were engaged in a fight against the Devil, just as the missionaries who had converted their heathen ancestors in Europe more than 1,000 years before. Every Catholic schoolchild knew the story of how Saint Boniface had cut down Donar or Thor's oak, which was sacred to the Germanic heathens. When the saint was not smitten dead because of this sacrilegious act, the surprised Germans who were present converted to Christianity. To the Catholic priests in Suriname, the kankantri was, like Donar's oak, a seat of the Devil. On passing a kankantri in 1842, the priest Peerke Donders (1809–1887) wrote: '[the tree] is remarkable, because the Devil receives homage there; because the tree is worshipped as a god by the Indians; here they come to sacrifice their chicken's eggs'.[14] In Africa, the tree was regarded as a spirit dwelling place. Slaves took this belief over the ocean and recognized the tree in the Caribbean. In the eighteenth and early nineteenth centuries, Dutch observers described how on certain nights the slaves performed their own secret rituals under the tree.[15] The slaves 'did not wish to fell this tree, however necessary, because of its aged roots. To persuade them to cut down the tree, one must give them drink … otherwise they will certainly protest against the whites'.[16] Stories abound in Suriname about the terrible accidents and serious diseases caused by cutting down kankantri trees.[17]

By raising a cross on a site of Devil worship, Van der Weijden intended to demonstrate his Christian god's superiority and thus European dominance over African slaves and their spirits. The missionaries who were sent out from the Netherlands to Suriname were all young and fresh from divinity college. They were unacquainted with Surinamese slave society and Afro-Surinamese culture and religion.[18] Suffering from the tropical climate, they were suffused with a spirit of sacrifice. This made them the ideal agents for instilling a 'Christian leper identity' in their congregation in Batavia, but gave them little insight into the distinct cultural mind-sets of the sufferers. In her study of Christian dealings with leprosy in colonial Africa, Megan Vaughan has called attention to 'the attempt to engineer socially a "leper identity" in the particular circumstances of colonialism.'[19] This included a projection of a Christian symbolism on the disease. Instead of heathen spirits or the breaking of heathen taboos taking a central place in a sufferer's understanding of leprosy, Christian religious practices and ideas were asserted. The Batavian asylum leprosy sufferers had to shed their former heathen cultural identity and forge a new Christian identity in order to receive the care of Catholic compassion.

Ironically, Van der Weijden did not survive the felling of the kankantri for long. After his return to Paramaribo, in October 1826, he came down with a severe fever and died after a week. However, he had cemented the symbiotic relationship between the Catholic Church and Surinamese colonial state. Jacobus Grooff (1800–1852), who was born in Amsterdam and consecrated as a priest in 1825 was sent to Suriname and furthered Van der Weijden's work. In 1827, Grooff was appointed as the apostolic prefect in charge of the mission and in the same year, he went to Batavia for the first time.[20] For Grooff, it was essential to give the solace of the Catholic religion to his 'poor sheep' in Batavia.[21] For instance, he washed the feet of twelve leprosy sufferers in Batavia in 1840 in a conscious imitation of Christ.[22]

Batavia in 1827

In August 1827, Andries van Hasselaar visited the Batavia asylum as a member of a committee of investigation. There had been rumours that a large number of healthy Afro-Surinamese were being kept in Batavia against their will. Governor De Veer decided to send a committee to investigate, which included Van Hasselaar and the city physician of Paramaribo (both official examiners of leprosy sufferers and members of the Collegium Medicum), his own military aide-de-camp, and a counsellor of the Court of Police and Criminal Justice. When the city physician cancelled his journey at the last moment, he was replaced by the naval surgeon of the corvette that took the committee from Paramaribo to the mouth of the Coppename River.[23] This left Van Hasselaar as the only leprosy expert. He was fairly upset by his experiences in the 'Empire of Lamentations' as he called the Batavia asylum, and so much so that the next year he went on sick leave to the Netherlands and never returned to Suriname.

Unlike Van der Weijden, Van Hasselaar's first impression of the Batavia asylum was positive. At first sight, Van Hasselaar found the asylum to be orderly, clean, and according to him unlike the usual 'dirt-heaps' in which the Afro-Surinamese lived. Landing on the riverbank, the committee members first passed the director's large wooden house and then the village street. Huts built of pineapple leaves lined both sides of the street and at the end of the street there was a small place for prayer. There was a large field with banana trees and a few smaller fields on which inhabitants who could still work cultivated karo, a kind of wheat. The wheat was used to feed the fowl and ducks that were traded with Amerindians and possibly even with slaves on nearby plantations, thus violating the segregation policy. Segregation was so incomplete that when the committee tried to examine leprosy sufferers it was told that an unspecified number were out on the river with the director's permission. Although the total number of leprosy sufferers in Batavia was unknown at the time, the committee examined approximately two hundred. A few had no signs of leprosy or elephantiasis. Some people were only living in Batavia because they had children afflicted with leprosy, and parents were not separated from their children. The committee chose an easy route and diagnosed all 200 sufferers as infected; for example, they included a man who was healthy but was living with a woman who showed symptoms of leprosy. Later, the surgeons explained that though ten out of 200 sufferers did not show signs of the disease, they were probably carriers of the leprosy poison anyway.[24]

Van Hasselaar and the other committee members' positive impression of Batavia radically changed when they entered the huts of sufferers who were not able to come outside for the examination. The malodour in the huts was so intense that some committee members went outside to vomit. The skin of some sufferers was so withered that they resembled the bark of an old willow tree. The hands and feet of other suffers had withered away and fallen off. Some sufferers had become completely crooked from their chin down to their knees. Still others had hoarse voices and their speech could not be understood. Judging from Van Hasselaar's account, faced with the horrors they observed, the committee members left Batavia as fast as possible since they were so upset by what they had witnessed.[25] The account shows that living conditions in Batavia were far from satisfactory. The committee's report confirmed the necessity of the Catholic missionary presence in Batavia to provide some form of sustenance to the sufferers.

The danger of Protestant competition

The leader of the Catholic mission, Grooff, visited Batavia for the first time in November 1827, baptized a further thirty leprosy sufferers and consecrated a new Catholic burial ground.[26] The alliance of the Catholics with the colonial state in the asylum seemed firm. However, there was one religious group among the Europeans in Suriname who objected to this development – the Protestant Moravians.[27]

The Moravians were also considered outsiders by the colonial establishment. The Moravian Church was founded in Switzerland in 1722 uniting various Protestant minorities including refugees from Moravia (hence the name) in an Evangelical Brother Community. The Moravians focused on a practical and non-dogmatic Christianity. Their founder, the Saxon Count von Zinzendorf, established a community called Herrnhut on his Swiss estate ('Herrnhutters' was a Dutch name for the Moravians). From the start, the community set up missionary activities overseas including in the Danish West Indies and from 1735 onwards in Suriname. In 1769, the Moravians in Suriname received permission to baptize and preach to the slaves on the plantations, which was more than sixty years before the Catholics had permission to do so.[28] For both religious groups, slaves were primary targets for conversion, so the Moravians did not relish Catholic competition after 1816. While the Catholics cemented their position in Batavia, the Moravians supported the treatment of leprosy sufferers in the aforementioned hospital established in 1828 by the homeopath Dr Hering. They also complained that the Catholics were baptizing slaves in Batavia who had already been baptized by the Moravians. One of these slaves told a Moravian missionary: 'what could I do [otherwise], so on my own?'[29]

Paulus Cantz'laar became Governor General of the Dutch West Indies in 1828. At first he tried to establish a clear demarcation line between the activities of the rival missions. In January 1830, he explicitly limited the Catholic's activities to Batavia and continued to deny them access to the plantations that were Moravian terrain.[30] At that time, there were 130 Moravian missionaries in Suriname and only seven Catholic missionaries.[31] Permission to be in Batavia was therefore vital for the Catholic mission. Grooff initiated the building of a new church and rectory, both of which were completed in 1836.[32] In 1832, he made an important breakthrough when the Catholics were given permission to extend their missionary activities to the plantations. Two years later, a new Governor General, the Baron van Heekeren, gave the land on which the church had been built in 'allodial possession' to the Catholics; that is, they received the right to use the land. Van Heekeren visited Batavia in 1835 and 1837, and his successor Jules Rijk followed with visits in 1839 and 1840. Both were quite content with the situation in the asylum.[33] The final recognition of the alliance between the colonial government and the Roman Catholic Church was Grooff's being awarded a knighthood in the Order of the Dutch Lion, one of the highest civic orders in the Netherlands, in October 1836. According to the royal proclamation, the knighthood was awarded for Grooff's services 'to those unhappy ones suffering from the contagious disease called Leprosy or Boasie and nursed in the asylum Batavia on the Coppename River'.[34] For its activities in Batavia, the colonial government gave the Roman Catholic Church a financial compensation of 1,000 guilders per year. In 1852, when a priest came to live full-time in the Batavia asylum, this compensation was raised to 1,500 guilders.[35] Occasionally, the sufferers would receive additional gifts of clothing and food from the Catholics financed by collections in the Netherlands.[36]

The support of the colonial state for Catholic activities in Batavia was a recognition of their essential role in maintaining order in the asylum. The Moravians remained the only possible outside threat to this influence. However, Moravian attempts over the next decades to preach in Batavia and establish a following there were unsuccessful.[37]

Population numbers

A survey of Batavia was held in 1833 and counted 401 segregated sufferers in the asylum. Exact figures for following years are only available from 1849 onwards when the numbers had to be entered in the government's Colonial Reports to Parliament in the Netherlands.

Batavia's population grew considerably after the new leprosy edict of 1830 came into effect and the machinery for detecting and segregating sufferers picked up speed. If all of the 194 persons who were declared as infected in 1831 had been sent to Batavia, and the fifty-three who were declared infected in 1832, the population must have doubled in two years. The count of 401 sufferers in 1833 means that Batavia was already overcrowded.

The 1833 survey also shows that Batavia was essentially an asylum for slaves. Only four of the sufferers were freed people, who were of a low socio-economic status and lacked a family network to support them outside the asylum. One of the four was a European and two were mulattos. Most remarkable is that seventy-six of the inmates, including forty-four children, were not infected – almost one in five. More than half of the children had been born in Batavia. No one in the colonial government appeared to have entertained the notion that these people had to be set free. Either their families were also in Batavia, or they were presumed to be carriers of the leprosy poison who had not yet developed the disease.[38] In 1849, the number of non-infected person living in Batavia had risen to 104, or around 20 per cent.[39] At the same time, the number of freed people in the asylum rose during the period of the 'Great Confinement' policy. In 1849, there were twenty-one freed people out of a total of 498 segregated sufferers in Batavia, which represented only slightly more than 4 per cent. In 1852, this number reached 6 per cent (27 of a total of 454 segregated sufferers) only to dwindle again to 4 per cent on the eve of emancipation in 1862 (15 of 362 segregated sufferers). The number of non-infected persons in Batavia was 89 in 1863, almost a fourth of the total number of the segregated population. The figures in the Colonial Reports to the Dutch Parliament after 1848 show a slight decrease in the total population, especially from 1854 onwards when the 'Great Confinement' policy was less stringently applied. After the emancipation of the slaves, the downward trend continued, reflecting the continual decline of segregation policies through the end of the century. In 1864, the year after emancipation, the number of segregated sufferers fell below 300, and at the end of the decade to below 200. From 1884 onwards, the number of segregated sufferers remained below 100 until the asylum was closed in 1897.[40] (See Figure 2 and Table 2.)

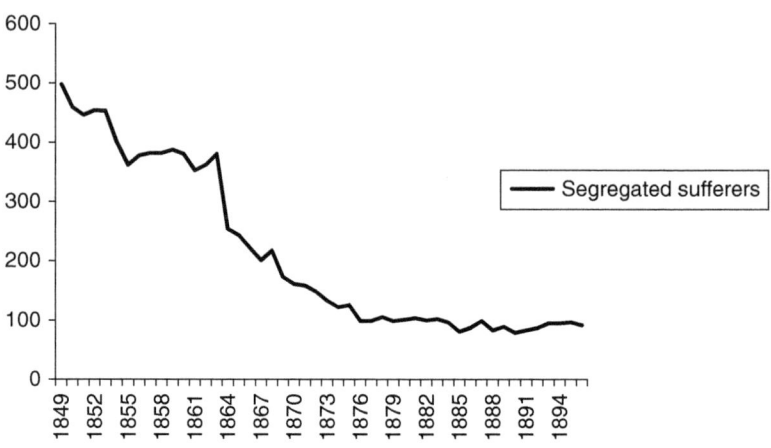

Figure 2 Segregated sufferers in Batavia, 1849–1897

Table 2 Segregated sufferers in Batavia, 1849–1896

Year (31 December)	Segregated sufferers
1849	498
1850	459
1851	446
1852	454
1853	453
1854	400
1855	362
1856	378
1857	382
1858	382
1859	387
1860	380
1861	353
1862	362
1863	380
1864	254
1865	243

Table 2 (*cont.*)

Year (31 December)	Segregated sufferers
1866	222
1867	201
1868	217
1869	173
1870	161
1871	158
1872	148
1873	133
1874	122
1875	125
1876	99
1877	99
1878	105
1879	99
1880	101
1881	104
1882	100
1883	102
1884	96
1885	81
1886	88
1887	99
1888	83
1889	89
1890	79
1891	83
1892	87
1893	95
1894	95
1895	97
1896	92

Source: Colonial Reports.

Order and disorder

Everyday life in Batavia must have been a crowded experience in the age of slavery. Did the Catholics succeed in imposing a religious disciplinary regime in the asylum? In his *Geschiedenis van Suriname* ('History of Suriname') (1861), chronicler Julien Wolbers expressed his doubts. The Catholic mission in Batavia had flourished under Grooff's leadership and 330 of the 353 sufferers were Catholic. However, although the priests showed a certain diligence in their work, according to Wolbers, not all priests conducted strict surveillance of their flock's moral behaviour.[41] Was Wolbers, a Protestant who had never been in Suriname, correct? The image promoted in the Surinamese newspapers and Catholic press in the Netherlands portrayed another situation – a Batavia inhabited by a grateful and obedient flock.[42] In 1837, a report in a Surinamese newspaper on the Governor's visit to Batavia mentioned poor sufferers taking solace in the Gospel and thankfully waiting for the arrival of their protector while dressed in white. Unfortunately, a more detailed description of everyday life in Batavia was missing in the newspapers.[43] In letters to the Dutch Catholic press, Grooff described Batavia as his own 'garden of pleasure' where he loved to spend time with his 'poor infected sheep', the sufferers.[44] The Saint Rochus Church that was consecrated in 1836 later attracted believers from the surrounding region despite the supposed segregation of the sufferers. In addition, a chapel was built in Batavia in 1841 and Catholic sufferers went there three times a week in procession to offer their prayers.[45]

The following account from the first published biography of Peerke Donders was typical of the Catholic depiction of life in Batavia. Donders was a Dutch priest who arrived in Suriname in 1842 and who became the iconic figure of Catholic leprosy care – the Dutch version of Father Damien (a well-known Belgian Catholic priest who worked with leprosy sufferers in Hawaii in the 1870s–1880s). From 1858 until his death in 1887, Donders lived and worked (with the interruption of a few years) in Batavia and among the nearby Amerindians as a missionary.[46] On Donders' first visit to Batavia, in October 1842, he was accompanying Bishop Grooff. As their boat approached Batavia, the bishop had soldiers fire gunshots 'to sign to the lepers that their father arrived'. The church bell was rung, and neither Donders nor Grooff could hold back their tears when they saw the 'poor lepers' coming towards them.[47]

The account accentuated the work of the Catholic Church in saving the souls and caring for the bodies of the leprosy sufferers.

It was not until 1844, however, that a priest had a permanent residence in Batavia – twenty years after the foundation of the asylum, and six years more before the colonial government saw fit to appoint a resident medical doctor. The first to be appointed as the resident priest was the thirty-year-old Gerardus Johannes Heinink. Heinink combined his sacerdotal duties with his directorship of Batavia, thus completing Catholic power.[48] Although Heinink's activities were described in the Catholic press in the Netherlands as satisfactory, in reality they were problematic.[49] When a naval surgeon Friedman visited Batavia (probably in 1849), he described the priest as a 'pale man' who 'possibly had been banished to this place because of some offence'. To Friedman, Batavia was a 'place of lamentation and disaster'.[50] Friedman's observations were in contrast to the positive comments in the Colonial Report of 1849 describing the many improvements in Batavia introduced by the colonial government, the director, and Grooff.[51]

Heinink's resources were limited when it came to nursing the sufferers. Not all Batavian inmates were willing to submit to Catholic morals and discipline. According to the Catholics, the 'notorious negro', a slave named Andreia, was an agent of chaos. He had his own rowing boat in which he frequently left Batavia without permission, and ran his own rum-smuggling racket, delivering alcohol to the inmates of the asylum in defiance of Catholic supervision. Reprimands were of no avail. In October 1849, Heinink caught Andreia in the act of smuggling and took more drastic action by sinking the slave's boat in the river. That same evening the priest became ill after supper, and by the next day he was dead. What shocked the Catholics most profoundly was that Heinink had died without receiving last sacraments. An investigation was started, but it was only after Batavia was put under interdict and refused Catholic services that witnesses came forward. A witness informed on Andreia who had had Heinink's milk poisoned by the priest's Afro-Surinamese housemaid. What happened to the two slaves afterwards was not recorded.[52]

Unrest continued among the inhabitants of Batavia, but their situation was improved a little by the colonial government's decision to appoint a resident physician in the asylum in 1850. The first physician Deutschbein died within a few months. His successor, Ooijkaas, arrived in 1853. Ooijkaas was quite negative about the health situation. Batavia was close to swamps and the sea, which meant that the scorching hot daytime air alternated with cold and foggy nights and damp mornings. Given prevailing notions about the connections between climate and disease, these changing temperatures were seen as important health risks. Furthermore, the sufferers' dwellings were reported to be in bad repair. Food allowances were meagre because of high costs. Fresh drinking water was scarce owing to a lack of waterholes. People had no woollen blankets for staying warm at night. Ooijkaas found that apart from the incidence of leprosy and elephantiasis, half of the inmates suffered from remittent fevers, and one in fifteen suffered from some sort of infection or rheumatic pain. The doctor's 'hospital' consisted of two small rooms, each with a tiny window to let in some air and light, but which were barred because the rooms were also used as a courtroom and jail. The doctor complained that he could not give adequate treatment in these circumstances. On average, four people died each month.[53] Ooijkaas died in 1855, and under his successor Carl Uhlig (a physician working from 1856 to 1863), there was a small improvement in the mortality figures. However, there are no indications that the living conditions in Batavia significantly improved during this time period.[54]

Meanwhile, Heinink's successor as resident priest in Batavia C. J. Magnée tried to impose discipline on the sufferers, an attempt that was considered too strict by both his colleagues and the inmates. Magnée lived in discord with the sufferers and excommunicated one of them, although this did not mean that the sufferer had to leave Batavia.[55] Unrest became so great that the discord was mentioned in the Colonial Reports of 1854 and 1855.[56]

Batavian unrest started to threaten the symbiotic relationship between the colonial state and the Roman Catholic Church. Faced with the breakdown in order in the asylum, the Moravians appeared to have a chance to become established in Batavia. They had a missionary present on a plantation that was a one-day journey from Batavia, and the wife of the then director of Batavia Van der Hoop was a Moravian.[57] This woman invited the Moravian preacher Van Calker to hold sermons in Batavia.[58] Grooff, who had had good contacts with the colonial government, had died in 1852 and the new head of the church in Suriname, Jacobus Gerardus Schepers, complained to the governor, but received a cold shoulder, possibly because of the governor's dissatisfaction with the results of Catholic activities in the asylum.[59] In 1857, Van Calker visited Batavia nine times to preach.[60] However, the odds were stacked against the Moravian since there were only seventeen Protestants in Batavia compared to almost 350 Catholics.[61] Magnée died in 1856 and Peerke Donders came to Batavia to clear up his mess and save the symbiotic relationship between the Catholics and the colonial state.

According to the Catholics, when Donders arrived in Batavia he found the moral standard of the inmates extremely low. Drunkenness and sexual debauches were believed to be the order of the day.[62] It was even suggested in the Dutch Parliament that the sexes in Batavia should be segregated to prevent intercourse between the inmates.[63] To restore order, Donders developed a two-sided strategy that he would continue to execute over the next thirty years. On the one hand he was strict with the inmates when enforcing discipline. He verbally abused them when he was dissatisfied with their behaviour and reprimanded them in sermons and prayers.[64] Just as his predecessors had done, Donders took up the fight against Afro-Surinamese folk beliefs and Winti practices that he regarded as Devil worship and black magic. He countered these practices with his own Catholic magic. He made sure there always was a supply of holy water in all dwellings. Donders often treated open wounds by prayer while dressing the wound with this holy water. It was said that in this way he cured an Amerindian who had been bitten by a snake. He let the members of his congregation wear scapulars at religious services and taught them how to make the sign of the Cross, thus expelling evil spirits.[65]

On the other hand, Donders often took the sufferers' side against the Protestant director of Batavia, Van der Hoop. Van der Hoop was reputed to freely hand out punishment for violations of the rules.[66] The 1830 regulations for Batavia entitled the director to arrest inhabitants and lock them up in jail (the two little rooms that also served as the doctor's hospital) for a period of eight days in chains or without according to his discretion. Furthermore, the director had the power to extend the period of arrest.[67] Donders gained credit among the inhabitants by taking their side and asking the director for forgiveness. This strengthened the Catholics' position and made the director ultimately responsible for anything that was unsatisfactory.[68] Later, Donders would successfully lobby for the replacement of earthen floors in the huts by wooden ones and a distribution of beds for the sufferers.[69] Thus, Donders succeeded in eliminating Protestant competition. In 1861, a Moravian missionary publicly gave up hope on his activities in Batavia.[70]

However, Donders was never completely successful in restoring order in Batavia. The accounts of Catholic witnesses also included stories of Donders' conflicts with inhabitants, such as the occasion when leprosy sufferers smuggled in rum from a nearby plantation and became inebriated and rowdy. The priest tried to calm them down but was cursed and even physically assaulted by the intoxicated.[71] The healthy rowers of the government boats and the healthy servants 'abused' (sexually?) the leprosy sufferers and held nightly dance parties with them. If Donders appeared on the scene of the party and told them to stop, he was cursed and driven away.[72] Government investigations revealed that sexual licentiousness was still rife in Batavia on the eve of the emancipation of the slaves. The Catholics reacted by forcing the inmates to marry. However, church marriages did not have legal status in Suriname. After emancipation, the colonial government did not recognize the validity of the marriages in Batavia. The marriages were broken up by sending healthy partners to other plantations for a period of 'state custody', which included a transition period of ten years in which the former colonial slaves remained under government control and were allocated as labourers to various plantations. The sufferers' children were sent to Paramaribo, probably to their families.[73]

Conclusion

The Batavia leprosy asylum served to segregate the poor and the slaves among the Surinamese leprosy sufferers, thus countering the threat that the disease posed to colonial society and the slave economy. It was also the place where the colonial state and the Roman Catholic Church formed a symbiotic alliance, thus strengthening the fabric of colonial society and furthering the assimilation of the Catholic minority in the colony. This was possible because although segregating unproductive sufferers in Batavia was crucial to the 'Great Confinement' policies, the colonial government did not wish to spend resources on them. The overwhelming majority of the Batavia inhabitants were

Afro-Surinamese slaves. This explains both the slaves' allure as a coveted object of religious groups for conversion, and the lack of interest in improving living conditions in the asylum. After all, leprosy was framed as a disease of an inferior race and people with immoral sexual behaviour. Isolation in a village in the jungle was all that these sufferers could expect from their masters since these slaves had ceased to be of economic interest. In the age of slavery, Batavia essentially remained a dumping ground for leprosy and elephantiasis sufferers, and a last port with no recall and inadequate medical and social care. What the colonial government did do was enter into a working relationship with the Roman Catholic Church, in which the Church had an essential role in maintaining some kind of moral and social order in the asylum. The Church was also allowed to use the asylum as an important recruiting ground for new converts and a base for missionary activities in the surrounding region. In the asylum, Catholic priests attempted to establish a 'Christian leper identity' on the sufferers, who had to shed their former heathen cultural identity to receive the care of Catholic compassion. But not all of the sufferers went along with the establishment of this new identity: Batavia was a place with its own infrapolitics of noncompliance and contestation that was not so evident in open resistance as in a cherishing of autonomy in everyday life. Batavia remained a contested battleground in the jungle.

Notes

1 O. ten Hove, '19e eeuws bevolkingsonderzoek naar lepra in Suriname', *OSO* 22 (2003), pp. 34–49; E. Klinkers, 'De bannelingen van Batavia. Leprabestrijding gedurende de negentiende eeuw in koloniaal Suriname', *OSO* 22 (2003), pp. 50–61. J. Vernooij, 'Een opvallende relatie. De rooms-katholieke kerk en lepra in Suriname', *OSO* 22 (2003), pp. 62–8.
2 M. D. Teenstra, *De landbouw in de kolonie Suriname* (Groningen: Eekhoff, 1835), pp. 156–7.
3 Archive Bisdom of Paramaribo (hereafter BP): T1, 'Kronijken van het etablissement Batavia', p. 3.
4 A. van Hasselaar, *Beschrijving der in de kolonie Suriname voorkomende elephantiasis en lepra (melaatschheid)* (Amsterdam: S. de Greber, 1835), p. 48, describes a voyage from Paramaribo to Batavia.

5   C. L. Drognat Landré, *De besmettelijkheid der lepra arabum, bewezen door de geschiedenis dezer ziekte in Suriname* (Utrecht: J.L. Beijers, 1867), p. 23, based on a report to the Court of Police.
6   BP: T1, 'Kronijken', p. 3.
7   A. Bosser, *Beknopte geschiedenis der katholieke missie in Suriname* (Gulpen: M. Alberts, 1884), p. 76.
8   On the history of the Roman Catholic Church in Suriname: M. F. Abbenhuis, 'De katholieke kerk in Suriname', *Vox Guyanae* 2 (1956), pp. 117–44; J. Vernooij, *De rooms-katholieke gemeente in Suriname. Handboek van de geschiedenis van de Rooms-Katholieke Kerk in Suriname* (Paramaribo: Leo Victor, 1998); see also Vermooij, 'Opvallende relatie'. For the Catholic mission in Suriname: A. J. J. M. van den Eerenbeemt, *De missie-actie in Nederland (±1600–1940)* (Nijmegen: J. J. Berkhout, 1946), pp. 50–4.
9   BP: T1, 'Kronijken', p. 3; Bosser, *Beknopte geschiedenis*, p. 178.
10 Bosser, *Beknopte geschiedenis*, pp. 170–2.
11 K. Bijker, 'Power, prayer and colonial pacification: The Roman Catholic mission in nineteenth century Surinam', in M. Bax and A. Koster (eds.), *Power and Prayer: Religious and Political Processes in Past and Present* (Amsterdam: VU University Press, 1993), pp. 57–79.
12 BP: T1, 'Kronijken'.
13 Bosser, *Beknopte geschiedenis*, pp. 195–6.
14 Cited in J. A. F. Kronenburg, *De eerbiedw, dienaar Gods Petrus Donders C.ss.R. Nieuwe levensbeschrijving* (Tilburg: W. Bergmans, 1925), p. 77.
15 A. Blom, *Verhandeling over de landbouw in de colonie Suriname* (Amsterdam: J.W. Smit, 1787), p. 346; P. J. Benoit, *Reis door Suriname. Beschrijving van de Nederlandse bezittingen in Guyana* (Zutphen: De Walburg Pers, 1980), pp. 64–5.
16 Lemmens, A. F., *Bijdragen tot de Kennis van de Kolonie Suriname. Tijdvak 1816–1822* (Geografisch en Planologisch Instituut VU Amsterdam, 1982), p. 180.
17 T. van Andel and S. Ruysschaert, *Medicinale en religieuze planten van Suriname* (Amsterdam: KIT, 2011), pp. 305.
18 Bosser, *Beknopte geschiedenis*; Vernooij, 'Rooms-katholieke missie'; Vernooij, *Rooms-katholieke gemeente*.
19 M. Vaughan, M., *Curing their Ills: Colonial Power and African Illness* (Cambridge: Polity Press, 1991), p. 77.
20 A. C. Schalken, 'Historische gids bestaande uit chronologische lijst naamlijsten varia registers. 300 jaar R.K.-gemeente in Suriname 1683–1983'

(Paramaribo, 1985), p. 14. Letters of Grooff, in which he describes his experiences in Batavia, were published in the Catholic magazine *De Godsdienstvriend* (hereafter GV); e.g., 34 (1835), pp. 35–54.
21 Bosser, *Beknopte geschiedenis*, p. 220. For Grooff, see J. Kleijntjes, 'Mgr. Grooff, apostolisch vicaris van Batavia', *Bijdragen voor de Geschiedenis van het Bisdom van Haarlem* 47 (1931), pp. 373–468; Vernooij, 'De rooms-katholieke missie'.
22 Bosser, *Beknopte geschiedenis*, pp. 222–4.
23 Details about the medical doctors and surgeons working in Suriname at the time: *Surinaamsche Almanak voor 1828* (Paramaribo: Maatschappij tot Nut van 't Algemeen, 1827).
24 Drognat Landré, *Besmettelijkheid der lepra*, p. 25, mentions the report to the Court of Police and Criminal Justice.
25 Van Hasselaar, *Beschrijving*, pp. 48–63.
26 Schalken, 'Historische gids', p. 14.
27 M. Lenders, *Strijders voor het lam. Leven en werk van Herrnhutter broeders en zusters in Suriname, 1735–1900* (Leiden: KITLV Press, 1996).
28 Bijker, 'Power, prayer and colonial pacification', p. 59.
29 Klinkers, 'Bannelingen', p. 58.
30 Schalken, 'Historische gids', pp. 14–15.
31 Bijker, 'Power, prayer and colonial pacification', p. 64.
32 [An.], 'Inwijding der kerk op het etablissement *Batavia* kolonie *Suriname*', GV (1836), pp. 247–53.
33 Schalken, 'Historische gids', pp. 15–16.
34 SN: *Curacaosche Courant*, 25 February 1837; www.delpher.nl/nl/kranten/ (hereafter: SN).
35 Bijker, 'Power, prayer and colonial pacification', p. 66.
36 Bosser, *Beknopte geschiedenis*, p. 222; Klinkers, 'Bannelingen', p. 53.
37 See Klinkers, 'Bannelingen', p. 58. For Catholic comments on such attempts in 1842, see GV 49 (1842), pp. 16–28.
38 National Archive, The Hague, Algemene secretarie Nederlandsche West-Indische bezittingen in Suriname 1830–1847, nr. 12.
39 'Mededeelingen nopens de *lepra* in onze West-Indische bezittingen', *Nieuw Praktisch Tijdschrift voor de Geneeskunde* 28, Nieuwe reeks 1 (1849), pp. 546–68, 761–70, on pp. 356, 770.
40 Colonial Reports (hereafter: CR).
41 J. Wolbers, *Geschiedenis van Suriname* (Amsterdam: De Hoogh, 1861), p. 767.
42 Letters of Grooff and other Catholic dignitaries in GV 43 (1839), pp. 69–75; 44 (1840), pp. 150–3; 45 (1840), pp. 169–74; 46 (1841), pp. 22–8; 49

(1842), pp. 16–28, 194–5; 50 (1843), pp. 50–3, 226–8; 54 (1845), pp. 50–5, 296–301; 61 (1848), pp. 163–78; 62 (1849), pp. 251–2.
43 SN: *Nieuwe Surinaamsche Courant* 25 October 1837.
44 Grooff, letter in *GV* 44 (1840), p. 152.
45 Bosser, *Beknopte geschiedenis*, p. 222.
46 In 1982, Donders' beatification took place in Rome and at the present time in the early twenty-first century, attempts by the Catholic Church in Suriname to elevate Donders to sainthood continue. Biographies of Donders, all from a Catholic apologetic perspective: F. Schweigman, *Twee Missionarissen onder de Melaatschen en Indianen van Suriname* (Roermond: J. J. Remen, 1894), pp. 1–173; F. Schweigman, *Aan de leden van het Hofbauer-liefdewerk: Pater Donders* (Amsterdam: Borg, 1900); Kronenburg, *Eerbiedw, dienaar Gods*; N. Govers, *Leven van den eerbiedwaardigen Petrus Donders C.ss.R. Apostel der indianen en melaatsen in Suriname* (Heerlen: Joh. Roosenboom, 1946); H. Helmers, *Een groot Nederlander in Suriname. Leven en werken van den eerbiedw. Dienaar Gods Petrus Donders* (Tilburg: Henri Bergmans, 1946); J. L. F. Dankelman, *Peerke Donders. Schering en inslag van zijn leven* (Hilversum: Gooi en Sticht, 1982).
47 *Twee Missionarissen*, p. 93; GV 50 (1843), pp. 50–3; Kronenburg, *Eerbiedw. dienaar Gods*, pp. 72–6.
48 Bosser, *Beknopte geschiedenis*, p, 231. The *Chronicles of Batavia* give his name as Heininck, but see his necrology in SN: *De Tijd* 2 February 1850.
49 GV 54 (1845), pp. 50–5; 296–300.
50 S. Friedman, *Nederlandsch Oost- en West-Indië, volgens de nieuwste inrigting, met betrekking tot aardrijkskunde, statistieken, voortbrengselen, luchtgesteldheid, en vooral tot den gezondheidstoestand* (Amsterdam: J. C. A. Sülpke, 1861), p. 222.
51 CR 1849, p. 315.
52 BP: T1, 'Kronijken', pp. 5–6; Bosser, *Beknopte geschiedenis*, p. 237; *Twee Missionarissen*, p. 95; Klinkers, 'Bannelingen', p. 51; SN: *Nieuwe Rotterdamsche Courant*, 14 February 1850.
53 Schneevoogt, V., 'Verslag op het rapport van den heer Ooijkaas, omtrent het lepreuzen etablissement Batavia, in de kolonie Suriname', *Verslagen en Mededeelingen der Koninklijke Akademie van Wetenschappen* 2 (1854), pp. 381–8.
54 Appointment Uhlig: CR 1856, p. 24; Resignation Uhlig: SN: *De West-Indiër*, 1 October 1863.
55 BP: T1, 'Kronijken', p. 6; *Twee Missionarissen*, p. 96.

56 CR 1854, p. 302; CR 1855, p. 28.
57 BP: T1, 'Kronijken', p. 6.
58 *Berigten uit de Heiden-Wereld* 23 (1857) pp. 58–64.
59 BP: O 28, letter Governor to Schepers, 29 August 1855; Bijker, 'Power, prayer and colonial pacification', p. 66; Klinkers, 'Bannelingen', p. 58.
60 CR 1857, p. 80.
61 Van Calker: *Berigten uit de Heiden-Wereld* 23 (1857) pp. 58–64; numbers of Catholics: CR 1857, p. 22. The numbers of Van Calker and the Colonial Report differ very slightly since they counted on different dates.
62 BP: T1, 'Kronijken'; *Twee Missionarissen*, pp. 96–8.
63 *Handelingen Staten-Generaal*: www.statengeneraaldigitaal.nl/, Kamerstuk Tweede Kamer 1855–1856, p. 491. Members of Dutch parliament expressed amazement at the high number of births in Batavia in 1860: SG: Kamerstuk 1862–1863, nr CII ondernummer 6, p. 1483.
64 An early Catholic evaluation of Donders' activities in Batavia, based on the Latin documents of his beatification trial after his death and especially on the witness statements of leprosy sufferers is Kronenburg, *Eerbiedw. dienaar Gods*, pp. 118–50. Cf, also BP: T1, 'Kronijken'.
65 Kronenburg, *Eerbiedw. dienaar Gods*, pp. 206–7.
66 Kronenburg, *Eerbiedw. dienaar Gods*, p. 126.
67 BP: T (no inventory number on document).
68 Kronenburg, *Eerbiedw, dienaar Gods*, p. 127.
69 Klinkers, 'Bannelingen', p. 53.
70 Klinkers, 'Bannelingen', pp. 58–9.
71 Kronenburg, *Eerbiedw. dienaar Gods*, p. 127.
72 Kronenburg, *Eerbiedw. dienaar Gods*, p. 208.
73 BP: T1, 'Kronijken', pp. 11–12.

PART II
LEPROSY IN A MODERN COLONIAL STATE

5

Transformations and discussion: Suriname and the Netherlands, 1863–1890

After the emancipation of the slaves in 1863 in Suriname, the legacy of leprosy control and fear of the disease in the slave society continued to affect leprosy management until the end of direct Dutch rule in 1950. This situation continued despite the profound changes that slowly turned the colony into a 'modern' colonial state. These changes were related to the transition of the plantation economy to twentieth-century late colonial capitalism in which large-scale agriculture lost importance to the production and export of products such as bauxite, gold, balata, timber, and oil dominated by American and Dutch companies. Large-scale immigration of indentured labourers from British India and the Dutch East Indies replaced slave labour. The Dutch wished to assimilate all inhabitants in the new colonial society and educate them to citizenship. Compulsory education and repression of Afro-Surinamese religious practices and language were considered essential to assimilation. In the modern colonial state, Dutch authoritarianism combined with a belief in rational order, linear progress, and standardized conditions of knowledge. From this perspective, modern Dutch colonial policies can be understood as an attempt at social engineering and described as 'authoritarian modernist', which is a useful term for distinguishing between the pre- and post-emancipation colonial states.[1] Colonial health policies to improve the standard of living of the non-white colonized population became an important element in the modern ideology of the colonizers.[2] As part of these health policies, compulsory segregation policies received new impetus in the 1890s. Compulsory segregation still existed between 1863 and 1890, but execution was lax.

This chapter examines Dutch debates about leprosy between 1863 and 1890. The debates took place when the threat of a 'return' of leprosy to the Netherlands appeared to materialise. This was ironic, since Dutch policy makers and doctors had to call upon medical expertise from Suriname at the same time as the Europeans were having medical debates about the validity of a contagionist theory for leprosy. In the Dutch East Indies, with much less direct Dutch control over the population, the hereditarian view of leprosy was embraced. However, in the Dutch West Indies, the international shift in medical thought towards a hereditarian rather than a contagionist view of leprosy did not affect the principle of segregation. The discussions show that developments in Suriname remained autonomous and were not directed from the Netherlands.

New migrant labour

The population of Suriname increased from almost 53,000 on the eve of the emancipation to almost 69,000 in 1898, of which only 833 were Europeans.[3] This increase was caused by the immigration of a new labour force of indentured contract labourers to work the plantations. The socio-economic character of Suriname changed profoundly as the export value of its chief product, sugar and sugar's by-products, such as molasses and rum, declined, although not the total tonnage of production. Production shifted to modernized large-scale plantations. By the beginning of the twentieth century, there were only four sugar estates still in operation, as compared to eighty-seven in 1860. Another problem for agriculture was the lack of available labour. After emancipation, the former slaves had to make labour contracts with former employers or others. However, many freed slaves chose to settle on remote plots of land away from the control of the colonial state. Others left the plantations in the interior and drifted to Paramaribo. Craftsmen and domestic servants also migrated to the town. As Dutch-Surinamese sociologist Rudolf van Lier wrote: the result was 'after Abolition the labour output by the former slaves decreased. Small groups of former slaves travelling around the countryside and moving from one plantation to the next also made for a somewhat restless situation, which was detrimental to plantation discipline and labour.'[4] The solution to this labour problem was the importation of indentured contract labourers, especially from Asia. Initially, approximately two thousand labourers came from China, and a few hundred from both Madeira and Barbados. By 1870, this initial source of labour disappeared as a result of the Chinese government's decision to prohibit emigration under contract altogether, and the British and Portuguese closed Chinese harbours under their control in Hong Kong and Macao. Labourers were then recruited in British India (present-day India and Pakistan) and more than 34,000 British Indians migrated to Suriname between 1873 and 1916.

The large majority of the British Indians came from what was then called the United Provinces (Uttar Pradesh), Bengal, and West Bihar. From the beginning of this migration, the Surinamese colonial government considered the British Indians to be unruly and there were reports of violent incidents caused by British Indian disappointment with the situation in Suriname and on the plantations.[5] At first, the British Indians had a right to a free return to India after serving their contracts, but in 1895, they were allowed to stay in Suriname permanently. The colonial government thus turned labour immigration policy into population policy to stimulate the colony's development. Two-thirds of the British Indian migrants stayed in Suriname permanently and many became small farmers in the districts. Although they came from various parts of India, these migrants developed a syncretic language called Sarnami Hindoestani. Since they came from Hindu regions in British India, the Dutch called them either British Indians or 'Hindoestanen'. The large majority of these immigrants were indeed Hindu, but there was also a sizeable Muslim minority of 17.5 per cent. By 1946, the British Indians accounted for more than a quarter of the total population of Suriname.[6]

The next important source of labour was Java in the Dutch East Indies. The immigration of Javanese indentured labourers started in 1890 and continued until the outbreak of the Second World War. Of the 38,620 East Indians who came to Suriname, one-fifth returned to the East. As the British Indians had done, the Javanese initially came to work on the plantations. Later, they generally settled as small farmers. In the east part of Paramaribo, the Blauwgrond quarter became primarily a Javanese neighbourhood. A large majority of the migrants were Muslim, and a small minority were followers of traditional Javanese religion.[7]

In the 1880s, Colonial Reports began to mention sufferers of leprosy from new migrant groups living in Batavia. In 1885, four British Indians were living in Batavia, who upon arrival in Suriname had possibly been diagnosed with leprosy during their compulsory medical examination. By 1892, this number had increased to seventeen. The number of Chinese sufferers in Batavia was fourteen in 1887 and eleven in 1892. Javanese sufferers are, however, not mentioned in the 1890s.[8]

Reorganizations

The migrants decisively changed the socio-economic and cultural landscape of the colony.[9] The Dutch colonial government pursued a policy of assimilation for both the former slaves (and their descendants) and the new immigrant labourers. Emancipated slaves were granted citizenship, but these new Dutch 'citizens' barely spoke Dutch and had little Western-style education. In an analysis by Van Lier, Suriname emerges as an overseas territory with a mixed population that had insufficient contact with Western culture.[10] New citizens were not represented in the new Surinamese parliament introduced by the liberal Dutch government of the Netherlands in 1866, the Koloniale Staten (Colonial Estates). In the government's view, 'Suriname constitutes a settlement in the true sense of the word, an overseas Dutch territory the population of which must be granted all the privileges and liberties which … have through the ages been the heritage of every Netherlands citizen.'[11] However, the Colonial Estates' powers were limited. The Estates had the right to control the budget, but the final budget still had to be approved by the Dutch Parliament. The Governor had to function in accordance with royal decrees from the Netherlands. Elections in Suriname were based on census suffrage, just as in the Netherlands. Until the introduction of general suffrage in 1948, the number of qualified voters in Suriname never rose above approximately a thousand. Although the great majority of the population was not qualified to vote, in the vision of the Dutch government and Dutch colonial state, they still had to be educated in order to become true Dutch citizens. Compulsory education was introduced in 1876 and taught in Dutch, since African customs and culture had to be eradicated. Sranan Tongo or 'Negro-English', the language of the slaves, was banned from the schools. In 1874, African religious customs were labelled as idolatry and their practice constituted a criminal offence. Later, children of British Indian immigrants could be educated in special schools in their own language, but in 1906, these schools were closed and sacrificed to the principle of assimilation.[12] In this way, the liberalization of the Surinamese colonial society after 1863 prolonged and even intensified the cultural policies of its ancestor slave society, when the slave owners had fought against the survival of African culture.

The presence of the new migrant groups in Suriname led to changes in healthcare. In 1879, the organization of healthcare outside of Paramaribo included the newly created position of district physician ('districtsgeneesheer'). This was a response to the increasing presence of British Indian migrant labourers in the outlying districts. The district physicians were civil servants of the colonial state. The creation of the new position was part of a long-term reorganization of medical healthcare. In 1879, a medical inspector ('geneeskundig inspecteur') was directed to supervise the civil health service. The first inspector, Salomons, held the post from 1879 to 1908 and performed a key role in the reorganization of leprosy policies over this time. Healthcare for the poor was organized in the various neighbourhoods in Paramaribo. To lower the threshold to a medical career for children from the middle- and higher classes in Suriname in 1882, the medical school ('Geneeskundige School') was founded by the colonial government and opened its doors in Paramaribo. The school offered a five-year course for district physicians and pharmacists. Some course graduates went to the Netherlands to continue their medical education at a Dutch university. In 1896, a new medical committee ('geneeskundige commissie') took over the supervision of the civil healthcare system. Unlike its predecessor in the period of slavery (the CGOT), this committee consisted solely of healthcare professionals: the medical inspector (and chairman), three doctors, and one pharmacist. The new medical regulations of 1896 also stipulated the obligation of every physician to report cases of disease that threatened public health within twenty-four hours. The leprosy committee continued to exist and examine potential cases of leprosy.[13]

Impact on leprosy control

The social and economic transformations of colonial Suriname had an impact on the field of leprosy control policies. At first the social relationship that changed between the former slaves and former slave owners had a negative effect on the execution of leprosy detection and segregation policies. Former slave owners lost interest in providing healthcare for people who were no longer their slaves. During the time of slavery, district surgeons had travelled to the plantations to manage leprosy and other health problems and were compensated for their labour by the slave owners. Following emancipation, district medical care was restructured. No longer did slave owners invite and pay for the services of European medical practitioners in Suriname. Instead, the colonial government appointed medical officers in the districts and provided them with a salary and hospitals.[14] Henceforth, the patients had to go to the hospital for medical care, which had an adverse effect on disease detection. Leprosy politics were no longer intimately connected with slave labour management.

Although the fear of leprosy was still alive among the Dutch in the decade before emancipation, 1853–1863, the number of leprosy examinations and 'convictions' by the Committee of Investigations, admissions to Batavia, and the asylum population had all sharply diminished. Slave owners who hoped for financial compensation after emancipation had been reluctant to report slaves with leprosy to the Committee. After emancipation, the downward trend in these numbers continued. There are no figures available on the number of Committee examinations, but figures in the Colonial Reports show that although the population of Suriname as a whole increased, the population of Batavia decreased and remained stable at approximately a hundred from 1876 until the asylum closed in 1897. The detection and segregation apparatus of the colonial government was still functioning during this time, but with far less intensity than during the height of the 'Great Confinement' policies.

Calling upon expertise from Suriname: the Bronbeek affair

In the meantime, Surinamese medical knowledge of, and expertise on, leprosy, a heritage of slave holders' medicine, had to survive attacks from opponents in medical professions elsewhere in the Dutch empire. These attacks were paradoxically triggered by evidence that the fear of leprosy's 'return' to the Netherlands was finally justified and manifested in the Bronbeek affair.

Since the 1750s, Europeans living in Suriname or visiting the country had expressed anxiety that leprosy would return to the Netherlands if transmission of the infection among Africans and from Africans to Europeans was not controlled. As recently as 1850, this fear had found expression in advice from medical practitioners in Suriname to the Ministry of Colonial Affairs in the Netherlands. The reactions of medical doctors in the Netherlands showed that this fear was somewhat theoretical for the Dutch. Although not denying the possible danger, Dutch doctors had never actually seen patients with leprosy in the Netherlands since the disease had been more or less eliminated in the country since the seventeenth century.[15] Belief in the influence of environmental conditions on the aetiology of disease was still strong in European medicine, and the doctors' belief that the transmission of leprosy was far less likely under European conditions than in the tropical climate further stimulated their relative indifference to the possibility of a return of the disease. Paradoxically the efficiency of the Surinamese detection and segregation policies contributed to this indifference. Dutch and other European soldiers in Suriname when diagnosed with or suspected of leprosy were not allowed to return to Europe.[16] When the disease did reappear in the Netherlands, it did not originate from someone from the West Indies, but rather from the other side of the globe in the Dutch East Indies.

Leprosy's 'return' was first observed in Bronbeek, situated in the eastern Netherlands. Here in 1863, a home for invalid soldiers from the Dutch East Indies' armed forces was opened on a royal estate. From an early start, there were soldiers with signs of leprosy living in Bronbeek. However, although these soldiers slept in communal bedrooms with other soldiers, no one was particularly concerned.[17] In any case, the resident medical officer Nieuwenhuizen was not alarmed. He had served for years in the East Indies and had come to the conviction that leprosy was not contagious.[18] The Conservative Minister of Colonial Affairs, Nicolaas Trackanen, was not so indifferent. When visiting Bronbeek at the end of 1866, he was shocked to see leprosy sufferers.[19] On 2 January 1867, he warned the commander of Bronbeek about the 'evil' that could result from the residences of the sufferers in Bronbeek and argued that all means necessary should be used to segregate them.[20] The Minister's language was quite similar to the language used to talk about leprosy in Suriname. On 7 January, he ordered that the sufferers should ideally be segregated in a building separate from the other Bronbeek inmates. They could not be discharged from the asylum.[21] To his horror, the Minister then received word that one of the sufferers had been on leave in Amsterdam for several weeks. The Minister ordered that the sufferer's whereabouts should be immediately identified, and all the people with whom he had contact in Amsterdam be examined. Again, the Minister emphasized that all sufferers should be strictly segregated in the Bronbeek asylum.[22]

By mid-January 1867, the news of the leprosy sufferer's presence in Amsterdam leaked out to the public. On 11 January, the mayor of the nearby city of Arnhem, in the neighbourhood of Bronbeek, wrote to the commander complaining that the 'Asian plague' had broken out so close to his town.[23] An Arnhem daily newspaper then brought the news into the open.[24] In this politically sensitive situation, the Minister turned to the greatest available authority on leprosy he could find. He personally asked Charles Landré to participate in a medical investigation of the suspected leprosy sufferers in Bronbeek.[25] By contacting Landré, he turned to a man with far more experience with leprosy than any medical practitioner in the Netherlands. The Minister also showed that he had little faith in the claimed experience of an East Indies medical officer such as Nieuwenhuizen. Medical expertise from a relatively peripheral corner of the Dutch empire and a product of slave holders' medicine was thus called upon for a medical intervention in the centre of the empire.

Landré had migrated as a medical doctor to Suriname in 1840, at the age of thirty-four. In 1864, with his health in decline because of long exposure to the tropical climate, he finally retired, returned to Europe and settled in Paris.[26] In January 1867, he was again in the Netherlands and accepted the invitation of the Minister to assist in the Bronbeek investigation. On 22 January, he visited the asylum with a public health inspector and two members of the provincial medical council. These local physicians had no experience with and little knowledge of leprosy. The report on their investigations in Bronbeek makes it clear that Landré was the one who explained to the other physicians what the leprosy symptoms were, what his experiences in Suriname had been, and what the interpretation of their observations should be. In short, Landré directed their clinical gaze. Therefore, the investigation report is also an indication of how the Committee of Investigation in Paramaribo had performed its investigations over the previous decades.

Visible signs were the decisive cues for diagnosing leprosy during examinations in Bronbeek. The physicians examined ten men and from the beginning, Landré took the lead. He explained to the other doctors about the signs of leprosy. By then the differentiation between anaesthetic and tubercular leprosy was clearly established in medical literature and medical education.[27] The most obvious signs of leprosy were either nodes or tubercles on the body, or anaesthetic maculae or spots. However, in the examination report these signs were observed in the same patients at the same time, thus making the differentiation between anaesthetic and tubercular in actual practice somewhat academic. Seven of the ten soldiers that were investigated in Bronbeek displayed both tubercles and anaesthetic spots. The face of the first soldier examined, the forty-five-year-old Belgian, Jean Joseph Fourier, was infested with tubercles and he had anaesthetic spots. Landré named these psoriasis spots. Apart from these more obvious signs, Landré pointed to other signs that were typical of leprosy in his experience. Fourier and other soldiers' bottom eyelids were lowered and fixed. Landré also accorded great significance to the paralysis of the extensor muscles of the fingers. For instance, the fifty-two-year-old Dutch corporal Willem Theodoor Coops showed both of these signs, as did Fourier, although Coops had fewer tubercles. To establish whether his leprosy was more tubercular or more anaesthetic, Landré resorted to a method that had been used in Suriname since the mid-eighteenth century and had been taught to the doctor Fermin by an African slave woman. Landré stuck a needle in Coops' face and limbs to establish whether the spots were anaesthetic. Another sign that Landré pointed out was the presence of ulcers on the soles of the soldier's feet and especially under the big toes of two other soldiers.

Thus, Landré was instrumental and decisive in establishing a leprosy diagnosis for the ten investigated soldiers. The doctors declared that seven of the soldiers were suffering from leprosy or elephantiasis graecorum. No differentiation was made in the diagnosis between anaesthetic and tubercular leprosy. An eighth soldier, a fifty-five-year-old Dutch sergeant, Peter van Delen, did not display any tubercles or anaesthetic spots. However, he did have large psoriasis spots on his right arm and his back. Psoriasis as a specific disease was known in Europe, but doctors were not certain whether or not these spots could develop into leprosy. Therefore, Van Delen was declared a 'suspect' and it was decided that he should be investigated again after one year as was the practice in Suriname. The last two soldiers investigated did not show signs of leprosy.

The Minister's fears had come true since seven, possibly eight, soldiers in Bronbeek suffered from leprosy. However, he had an even bigger concern. He had asked the physicians to establish whether the disease had been contracted in the East Indies or in the Netherlands, or perhaps in Bronbeek. In other words, were people once again infected with leprosy in the Netherlands?

The medical history of the leprosy sufferers in Bronbeek was reason for concern. Both Fourier and Coops told the doctors that the signs of their disease had only started to occur *after* their return to the Netherlands. Fourier had returned five years earlier and had been admitted to Bronbeek in 1863, but his first signs of leprosy had only begun to appear after his admission, or so he claimed. Coops had also returned five years earlier, having been admitted to Bronbeek in 1863. According to Coops, the signs of his disease had first started to appear in 1865. The other five leprosy sufferers had already shown signs of leprosy when serving in the East Indies and had been admitted to Bronbeek between 1863 and 1865. Although theoretical, it was possible that the five had spread the infection to Fourier and Coops, but the doctors ruled this out. The doctors believed that Fourier and Coops' disease was too developed (both of them would die later in the year), and so the two soldiers must have been infected for longer than just three or four years. The doctors' conclusion challenged Fourier and Coops' claims of only having symptoms after arriving in the Netherlands. For instance, the doctors believed that Fourier had developed his psoriasis spots earlier than he had claimed, arguing that he had probably not noticed them or not recognized them as signs of the disease. Once again, preference was given to the expertise from Suriname and the physicians advised the Minister that there was no reason to suppose that any soldier had contracted leprosy in Bronbeek.[28]

Nevertheless, the leprosy poison was back in the Netherlands and precautions to prevent further transmission were called for. In February 1867, the Inspector General of the armed ground forces sent a 'very confidential' order to all medical officers to inform him without delay of any further cases of leprosy among the soldiers.[29] It took some time before the leprosy sufferers in Bronbeek were moved to another location. Both Coops and Fourier died in Bronbeek in 1867, the first in May and the latter in September. A third sufferer died in Bronbeek in July. The other soldiers, including the 'suspect' Van Delen, were officially discharged from the Bronbeek estate by their own request: three in March, and the last two in September.[30] There are no records of where the discharged soldiers went. However, in August 1867, the newspapers reported new cases of leprosy. Three leprosy sufferers were discovered during a visit to the penal institutes of Ommerschans and Veenhuizen in the eastern Netherlands, to which beggars, vagabonds, and poor orphans were compulsorily admitted. A fourth leprosy sufferer was found in the streets of Arnhem. He was sent to Veenhuizen, where a separate building had been constructed to house all leprosy sufferers and to segregate them from the other inmates.[31] There is no evidence to say whether any of the leprosy sufferers from Bronbeek were among the patients in Veenhuizen.[32] A sufferer and his healthy wife served as caretakers of the leprosy building, making it a kind of miniature Batavia.[33] The building received the rather grand name 's Rijks leprozenhuis ('State leprosy asylum').[34] In 1873, the number of sufferers in 's Rijks leprozenhuis had increased to eleven.[35] When the asylum was closed in 1886, there were fifteen or sixteen patients living there. All the patients had travelled to the East or West Indies. Fourteen had been admitted voluntarily, and one or two compulsorily as convicted vagabonds or beggars. The asylum closure had nothing to do with leprosy policies. A change in the law ruled that only persons capable of work could be admitted to Veenhuizen. The three last leprosy sufferers were unproductive and thus discharged.[36] Their fate was unrecorded. The closure of the asylum shows that the scare of the return of leprosy triggered by the Bronbeek affair had not lasted. By the time of the closure, the contagiousness of leprosy had become a contested subject of medical debate in the Netherlands.

The danger of contagion: the medical thesis of Drognat Landré

After the Landré family's return to Europe in 1864, son Charles Landré began his medical studies. In 1863, Charles had changed his last name to Drognat Landré to honour his brother Drognat who had died of leprosy in 1862.[37] In 1867, Drognat Landré observed two leprosy sufferers in the civilian and military hospitals in the town of Utrecht.[38] To father and son Landré, it was evident that although the fight against leprosy had been declining in Suriname, it had to receive new impetus in Europe. Leprosy became the subject of Drognat Landré's MD thesis, and he defended it at the University of Utrecht on 14 July 1867 when he was twenty-three years of age.

Drognat Landré's thesis on the contagiousness of leprosy was filled with both personal engagement and political relevance. Drognat Landré used his father's notes, archives and experience to make it clear to the medical world that leprosy as a contagious disease was still an important danger and should be fought with all means necessary. In doing so, Drognat Landré took a clear stance against the hereditary view of the Norwegians Danielssen and Boeck that was gathering increasing authority in international medical discussions.[39]

To Drognat Landré, a historical-epidemiological study of leprosy in Suriname provided more evidence and insight into the aetiology of the disease than studies in countries such as Norway, where the beginnings of leprosy's history were not documented. Therefore, the first forty pages of his thesis included a detailed account of the history of leprosy in Suriname. This was not so much historical background as essential evidence for Drognat Landré's argument. In the second half of his thesis, Drognat Landré summarized the experiences and conclusions of European medical practitioners and observers in Suriname. He wrote that leprosy did not exist among the Amerindians and was first observed among the Africans, who had carried it over the ocean to the Americas. Only later did Europeans become infected. Drognat Landré presumed that the main transmission mechanism of leprosy to Europeans was sexual contact between European males and African women. He believed that there was no evidence of leprosy in the families of these infected Europeans before this type of sexual contact. In Suriname, there was also no evidence to support the spontaneous generation of leprosy. Therefore, the disease must be transmitted by physical contact, and this meant the existence of a contagium, an agent capable of causing a communicable disease.[40] An insufficient diet did not cause leprosy, but could undermine individuals' resistance to the impact of the contagium.[41] To explain how not all inmates of the Batavia asylum had contracted the disease, Drognat Landré claimed that the impact of the contagium was in itself not so powerful.[42]

Drognat Landré transformed Schilling's leprosy poison into a contagium.[43] In the tradition of eighteenth-century colonial medicine, his argument included an emphasis on sexual customs. However, Drognat

Landré did update the Surinamese position on the contagiousness of leprosy by moving away from more fluid ideas about the respective influences on contagiousness of heredity, lifestyle, and climate that had characterized the eighteenth- and early nineteenth-century colonial medicine of Schilling and Van Hasselaar. In this way, he strengthened the contagionist argument against the views of Danielssen and Boeck. Claiming that he had embraced the opinions of medieval physicians, Drognat Landré concluded at the end of his dissertation that without contagium, there could be no leprosy.[44]

Contagion versus inheritance

By strengthening the medical argument for leprosy's contagiousness, father and son Landré took their fight against leprosy to Europe. In 1867, this was a rear-guard action, as can be seen from the medical discussions that were triggered by the publication of Drognat Landré's thesis. The Bronbeek and Veenhuizen affairs gave the leprosy discussion an acute and immediate relevance. At the same time, father and son Landré's Surinamese expertise lost authority in the eyes of Dutch doctors in the Netherlands owing to a change in international medical discussions of leprosy in which the decreasing importance of the Caribbean to the global colonial empires shone through.

Early in 1867, a committee of the British Royal College of Physicians issued a report on leprosy. Based on answers to a questionnaire by more than 250 medical practitioners around the world, the committee concluded that there was no evidence that leprosy was contagious and that reported cases of contagion were based on 'imperfect observation'. To the committee, leprosy was 'essentially a constitutional disorder' transmitted by inheritance. Strengthening the constitution by improving health, diet, and living conditions in native communities in the colonies was asserted to be the correct method of fighting leprosy. Compulsory segregation was rejected, although asylums were held to be useful for the prevention of the hereditary vulnerability to the disease by hindering sexual contact. The committee's report suggested that Europeans afflicted with leprosy in the colonies had usually been born there and grown up in tropical conditions.[45]

The committee's report gave new authority to Danielssen and Boeck's hereditarian view of the transmission of leprosy, but it ran counter to

Caribbean experiences and expertise. In places where British colonial administrators and doctors were directly confronted with leprosy, the Royal College report's conclusions were not generally accepted. When the committee secretary went to the British West Indies to investigate in 1871, he discovered that his report was almost unknown. He wrote that in British Guiana, Jamaica, Barbados, Trinidad, and other British colonies, the report 'had been seen by very few; most of the medical men even had not read it ... in more than one [colony], the government offices were without a copy'.[46] Developments surrounding leprosy in these British West Indian colonies were autonomous. Devoting too much weight and attention to discussions in medical journals of the nineteenth century obscures the actual practice of leprosy control in the British Caribbean. In British Guiana, Suriname's western neighbour, a leprosy asylum for infected slaves had been founded in 1832. The arrival of Chinese and Indian labour migrants since the early 1840s had increased the fear of leprosy and in 1858 a second asylum had been built specifically for the Chinese. In 1870, there were 303 segregated sufferers in the colony and the total number of sufferers was estimated at between 500 and 600, or 0.2 per cent of the total population. To a British doctor in Guiana, the Royal College's report was unconvincing and many of the cases mentioned in the report pointed to the role of contagion according to the doctor.[47]

Although the conclusions of the Royal College's 1867 report also ran counter to the views held in Suriname, they did receive support from the other side of the Dutch colonial empire in the Dutch East Indies. In 1865, the East Indies' colonial government had decided that leprosy was not transmitted by contagion, but rather by inheritance and therefore segregation was unnecessary. Both medical and financial reasons led to this decision.[48] Medical practitioners in the East Indies took a stance in the discussions on leprosy against their colleagues in the West Indies. The doctor at Bronbeek Nieuwenhuizen publicly stated his opinion that leprosy was not contagious.[49] Another doctor who had served with the military in the East Indies and opposed the contagion theory was Berend Carsten, a state medical inspector in the Netherlands.[50] Carsten had read Drognat Landré's thesis before the defence at the University of Utrecht and he wanted to calm the fears created by the Bronbeek affair in the influential *Dutch Journal of Medicine*. Carsten denied that there was evidence that people from the West or East Indies had infected people in the Netherlands. He cited the example of a patient in his private practice in Amsterdam who was a leprosy sufferer who had returned from the East Indies, and whose wife and children did not show symptoms of the disease. Carsten wrote extensively and approvingly of the conclusions of the Royal College.[51]

A third medical officer from the East Indies who publicly took sides against the contagion theory was Frederik Johannes van Leent. He was a naval health officer and lecturer at the naval hospital on the island of Curacao in the Dutch West Indies. Van Leent had served for an extensive period in the East Indies and had written his own MD thesis on the medical geography of the East Indies' capital Batavia (now Jakarta) in 1868.[52] The next year, he reviewed Drognat Landré's dissertation in the *Medical Journal for the Navy*. Though he did not totally deny the role of contagion, Van Leent claimed that there were no cases of contagion observed in the East Indies, and only one in the Navy in the West Indies. To Van Leent, inheritance was the most important transmission mechanism and he explained this with examples from the history of leprosy incorporating racial stereotypes in his analysis and thereby contributing to a further racializing of the disease. Not all members of a leprosy sufferer's family showed signs of the disease, he wrote, and this was not because they had not inherited it, but rather because of their strong constitutions. According to Van Leent, the Jews might have brought leprosy to Suriname and transmitted it by producing offspring with African slaves. Van Leent qualified the stigmatization of sufferers as a 'medieval curse' and compulsory segregation as 'inhumane'.[53]

Depending on their viewpoint, doctors could use the same data as evidence of inheritance or of contagion. In 1868, a publication in the *Dutch Journal of Medicine* described a forty-year-old female patient diagnosed with 'lepra Surinamensis' in Amsterdam. She had lived in Suriname as the wife of a non-commissioned officer whose brother and sister had leprosy and lived in a segregated dwelling behind the family home. Neither the patient's second husband in Amsterdam nor her parents were infected. Her physician concluded that the patient had been infected by her in-laws in Suriname, but that the disease was not communicable in the more northern climate of Amsterdam; hence he made his diagnosis of 'Surinamese leprosy'.[54] While this doctor adhered to older medical views, Van Leent claimed that the same case showed that there was no evidence of contagion, since the second husband was not infected.[55] Drognat Landré responded to his critics in the revised 1869 French-language edition of his published thesis and took the case as evidence of contagion; he suggested that the woman had been infected by her in-laws. Landré had investigated the woman (a mulatto) himself in Amsterdam. He claimed that mulattos did not have a fear of contagion or physical contact with leprosy sufferers.[56] To Drognat Landré sufferers living in Batavia provided further evidence of the contagiousness of leprosy. Van Leent disagreed, basing his opinion on a report by Uhlig who had worked as doctor in Batavia between 1856 and 1863.[57] Because the report stated that many inhabitants in Batavia were not infected, Van Leent thought that the disease could not be contagious.[58] Drognat Landré in turn attacked Uhlig and questioned his observations; for instance, the existence of non-infected European children of soldiers with leprosy.[59] From his father's notes, Drognat Landré added the case histories of twelve European children with leprosy in Suriname to the French edition of his thesis. He described them as 'full-blooded' Europeans with no traces of African or Jewish blood, born of wealthy parents and well fed. He presented these histories as evidence that the disease was not transmitted by inheritance. Among the case histories was a doctor's son who had contracted leprosy at the age of two and died at twelve.[60] Later, historians have shown that this was Drognat Landré's own brother Drognat, who had died in 1862.[61]

Under attack in the Netherlands, Drognat Landré and his father decided to take the contagion debate to an international level by publishing a French edition of Drognat Landré's thesis in Paris. In the Netherlands, medical opinion turned against contagionism. In 1868, H. J. Vinkhuijzen, a medical practitioner, published a Dutch monograph on leprosy.[62] Vinkhuijzen based his writing on 'the new scientific practice of leprosy' by Danielssen and Boeck, and by Virchow, and on Uhlig's observations. Vinkhuijzen decided against the theory of leprosy transmission by contagion as he considered leprosy to be a constitutional disease. He wrote that leprosy could develop spontaneously. Although inheritance could play a role in transmitting a predisposition to leprosy, he believed that lifestyle was the decisive factor. Vinkhuijzen wrote that one only had to witness the diet and dirtiness of the Africans.[63] He rejected Schilling's theory of contagion, but kept the latter's racist prejudices.

The general theoretical view in Dutch medicine seemed to turn against contagion and Surinamese expertise in leprosy. A month before Drognat Landré defended his thesis in Utrecht, another medical thesis on leprosy had been defended in Groningen. The author, C. H. Sanders, denied the role of contagion and accorded heredity and to some extent lifestyle and climate the most prominent roles in the transmission of leprosy.[64]

Medical debates without policy consequences

The debate in the Netherlands had few consequences for Suriname. The 'Great Confinement' policies had already been in decline since the 1850s and Drognat Landré's writings were directed as much against the new hereditarian framing of leprosy in Europe as against the perceived slackness of the execution of segregation policies in Suriname. The governmental and medical authorities in the Netherlands did not formulate an official opinion on the contagiousness of leprosy. In reports to Parliament, medical inspections showed some preference for Vinkhuijzen's position as opposed to Drognat Landré's.[65] However, this did not lead to the end of the segregation of leprosy sufferers in Veenhuizen. After all, the role of contagion was not the only argument for segregation; prevention of sexual reproduction was another. At the same time, this did not mean that the compulsory segregation policy in Suriname was completely unchallenged. If leprosy was not contagious, was there a need for a leprosy asylum in Batavia? Some found an opportunity here to economize in the Surinamese colonial budget.

The Royal College report had been discussed in the Surinamese press as early as March 1867 under the sensational heading, 'Leprosy is not contagious'.[66] Some members of the new Colonial Estates started to question the need for Batavia when discussing the Surinamese budget for 1867.[67] Governor Willem van Idsinga also expressed doubts about the future of Batavia.[68] He had only arrived at his post in Suriname early in 1867 and was probably more attuned to the discussions in Europe than to the Surinamese experiences in regard to leprosy. However, the fear of leprosy was too ingrained in Suriname to put an end to compulsory segregation. In October 1870, the Committee of Investigation in Paramaribo sent a report to the Minister of Colonial Affairs in the

Netherlands emphasizing that complete segregation of leprosy sufferers was essential in the fight against the disease. The Committee also advocated a strict separation of the sexes, and in doing so appealed equally to hereditarians. In discussions on the colonial budget for Suriname in 1871, some members of the Dutch Parliament entertained the idea that the leprosy asylum in Suriname could be closed just as in the East Indies and that this would improve the state of the colonial budget.[69] But the Minister refused, taking the advice of the Committee of Investigation seriously and also expressing his own doubts about the hereditarian argument. He knew of a recent case in the Netherlands that seemed to point to contagion.[70] In December 1873, the Minister of Colonial Affairs told the Dutch Parliament that although the population of Batavia had decreased slightly, one could not infer that the incidence of leprosy in Suriname was decreasing as well. According to the Minister, most leprosy sufferers were segregated in their own homes.[71] Batavia continued to exist.

In the Suriname of the 1870s and 1880s, there were no publications, studies, or public upheaval about leprosy. The hereditarian stance of the Dutch East Indies of 1865 was rejected in the West Indies, and thus the situation remained as status quo. Only Charles Landré (his son had moved to Brazil), who occasionally came out of retirement, warned about the contagiousness of leprosy. The discovery of the leprosy bacillus by the Norwegian Armauer Hansen in 1873 gave new ammunition for Landré's arguments, but the role of the bacillus in leprosy transmission was not generally accepted for some time. In 1883, in a congress on colonial medicine in Amsterdam, Landré (by then in his mid-seventies) suggested that the role of contagion could be proven by infecting pigs with the leprosy virus. He also stated once again that the leprosy policy in Suriname was defective and that a majority of the population and doctors there had to be convinced of the danger of contagion.[72]

A few years before his death, Landré published his last testimony on leprosy in a fifteen-page pamphlet in 1889. This was Landré's final rear-guard action in defence of the 'Great Confinement'. However, if one perseveres long enough in a rear-guard action, one can find oneself once again in the forefront. The discovery of the leprosy bacillus had changed medical opinions, and the contagionist view returned once again, Landré claimed.[73] The positive review of Landré's pamphlet in the *Dutch Journal of Medicine*, which was in complete agreement with Landré's viewpoints, showed that he was right.[74]

Conclusion

Leprosy politics in Suriname continued to be closely intertwined with the economic foundations of the colony and especially with labour transitions. The end of slavery put a temporary stop to the strict execution of confinement policies. However, the fear of the disease inherited from the slave society and slave holder's medicine was too engrained in Suriname to abandon views of contagionism and the necessity of compulsory segregation altogether. The international shift to hereditarianism passed Suriname by. Compulsory segregation policies in Suriname were still in place when the colonial government, medical practitioners, and religious groups once again felt the urgency of these policies and tried to reformulate them by 1890. The colonial state was then ready to give new impetus to the fight against leprosy. After decades, father and son Landré transformed from fossils of a past age into the forerunners of a new one. Compulsory segregation policies continued in Suriname in a modern colonial state and an economy based on indentured labour.

Notes

1 On authoritarian modernism, see J. C. Scott, *Seeing Like a State: How Certain Schemes to Improve the Human Condition Have Failed* (New Haven, CT: Yale University Press, 1998), pp. 88–97, 177.
2 D. Arnold, 'Introduction: Disease, medicine and empire', in D. Arnold (ed.), *Imperial Medicine and Indigenous Society* (Manchester: Manchester University Press, 1988), pp. 1–26.
3 Colonial Reports (hereafter CR) 1863, 1899.
4 R. van Lier, *Frontier Society: A Social Analysis of the History of Surinam* (The Hague: Martinus Nijhoff, 1971), p. 185.
5 C. J. M. de Klerk, *De immigratie van Hindoestanen in Suriname* (Amsterdam: Urbi et Orbi, 1953), pp. 139–48.
6 On the migration of British Indians to Suriname: de Klerk, *Immigratie*; M. S. Hassankhan, 'De immigratie en haar gevolgen voor de Surinaamse samenleving', in L. Gobardhan-Rambocus and M. S. Hassankhan (eds),

Immigratie en ontwikkeling. Emancipatieproces van contractanten (Paramaribo: Anton de Kom Universiteit, 1993), pp. 11–35; R. Hoefte, *In Place of Slavery: A Social History of British Indian and Javanese Laborers in Suriname* (Gainesville, FL: University Press of Florida, 1998); R. Bhagwanbali, *De nieuwe awatar van slavernij. Hindoestaanse migranten onder het indentured labour system naar Suriname, 1873–1916* (The Hague: Amrit, 2010).

7 On the Javanese: J. Ismael, 'De immigratie van Indonesiërs in Suriname' (PhD thesis, Leiden, 1949); A. H. de Waal Malefijt, 'The Javanese population of Surinam' (PhD thesis, Colombia University, 1964); P. Superlan, 'The Javanese in Surinam: Ethnicity in an ethnically plural society' (PhD thesis, University of Illinois, 1978); R. Hoefte, *De betovering verbroken. De migratie van Javanen naar Suriname en het rapport-Van Vleuten (1909)* (Dordrecht: Foris, 1990); Hassankhan, 'Immigratie'; Hoefte, *In Place of Slavery*.

8 CR 1886, 1888, 1893. The numbers are not very reliable, since in the following year the 95 sufferers in Batavia are classified as 91 Creoles (*inboorlingen*), two Chinese and two British Indians. CR 1894.

9 Van Lier, *Frontier Society*, pp. 180–246; H. Buddingh', *De geschiedenis van Suriname* (Amsterdam: Nieuw Amsterdam, 2012), pp. 217–26.

10 Van Lier, *Frontier Society*, p. 191.

11 Cit. in van Lier, *Frontier Society*, p. 190.

12 Van Lier, *Frontier Society*, pp. 191–3.

13 G.-J., Hallewas, 'De gezondheidszorg in Suriname' (PhD thesis, Groningen University, 1981), pp. 85–7, 95–6; *Surinaamsche Almanak voor het jaar 1888* (Paramaribo: Erve J. Morpurgo, 1887); district physicians and British Indians: de Klerk, *Immigratie*, p. 126; Bhagwanbali, *Nieuwe awatar*, pp. 118–38.

14 A. Kappler, *Surinam, sein Land, seine Natur, Bevölkerung und seine Kultur-Verhältnisse* (Stuttgart: J.G. Cotta, 1887), p. 187.

15 G. N. A. Ketting, 'Bijdrage tot de geschiedenis van de lepra in Nederland' (MD thesis, University of Amsterdam, 1922), p. 13.

16 See J. Karbaat, 'Sociaal-geneeskundige beschouwingen over de personeelsleden van de troepenmacht in Suriname en hun gezinnen' (MD thesis, Leiden University, 1963), p. 189.

17 *Gedenkboek van het koloniaal-militair invalidenhuis Bronbeek* (Amsterdam: P. Gouda Quint, 1881) contains sketches and maps of the asylum.

18 A letter of Nieuwenhuizen to the Dutch press is quoted in *De Tijd* 23 January 1867, www.delpher.nl/nl/kranten/ (hereafter SN).

19 At least, this is the impression one receives when reading a newspaper report circulating in Dutch newspapers, e.g., SN: *Algemeen Handelsblad* 17 January 1867, and the letter of the Minister to the commander of Bronsbeek, Archive Bronbeek, Arnhem (hereafter AB), 2005/00-160-79 8, 2 January 1867 (courtesy of Leo van Bergen).
20 AB Nr 2005/00-160-79 8, letter of the Minister to the commander of Bronsbeek, 2 January 1867.
21 AB 2005/00-160-79 8, letter of the Minister to the commander of Bronsbeek, 7 January 1867.
22 AB 2005/00-160-79, letters of the Minister to the commander of Bronsbeek, 8, 12, 16 and 18 January 1867.
23 AB 2005/00-160-79 8, letter of the Mayor of Arnhem to the commander of Bronsbeek, 11 January 1867.
24 SN: *Algemeen Handelsblad* 17 January 1867.
25 AB 2005/00-160-79, letters of the Minister to the commander of Bronsbeek, 21 January 1867.
26 H. E. Menke, 'The Landré Family: Drama and Scientific Concept in a Slave Colony Ravaged by Leprosy' (unpublished manuscript, Rotterdam, 2010).
27 For example, see R. Virchow, *Die krankhaften Geschwülste. Dreissig Vorlesungen* (Berlin: August Hirschwald), vol. 1, pp. 268-363.
28 AB 2005/00-160/79 8, report of the committee of investigation of leprosy in Bronbeek, 29 January 1867. Dates of admittance to Bronbeek and dates of death: *Gedenkboek van het koloniaal-militair invalidenhuis Bronbeek*, pp. 200-7.
29 L. van Bergen, 'De vreeselijkste van alle kwalen Lepra in Nederlands-Indië 1815-1942' (unpublished manuscript, Royal Netherlands Institute of Southeast Asian and Caribbean Studies, Leiden 2015), p. 91.
30 *Gedenkboek van het koloniaal-militair invalidenhuis Bronbeek*, pp. 200-7.
31 SN: Leydse Courant 7 January 1867; Provinciaalsche Ijsselsche en Zwolsche Courant, 16 August 1867; *Handelingen Staten-Generaal*: www.statengeneraaldigitaal.nl/ (hereafter SG): 'Begrooting voor de gestichten te Ommerschans en Veenhuizen', Kamerstuk Tweede Kamer 1868 kamerstuknummer 64 ondernummer 3, 4.
32 As stated in SG: 'Begrooting van uitgaven voor de Rijksgestichten Ommerschans en Veenhuizen', Kamerstuk Tweede Kamer 1875-1876 kamerstuknummer 109 ondernummer 107, p. 26.
33 SG: 'Verslag over de verrigtingen aangaande het armbestuur over 1867', p. 1189, Kamerstuk Tweede Kamer 1869-1870 kamerstuknummer 3 ondernummer 2.

34 SG: Handelingen Tweede Kamer 1875–1876, 2 May 1876.
35 SG: 'Verslag over de verrigtingen aangaande het armbestuur over 1873', p. 67, Kamerstuk Tweede Kamer onderstuknummer 67 ondernummer 2.
36 T. Broes van Dort, 'Een en ander over de lepra in Nederland en zijne koloniën', *Nederlandsch Tijdschrift voor Geneeskunde* pp. 292–6, 384–91, 407–21, 650–1, on pp. 650–1.
37 Menke, 'Landré family'.
38 C. L. Drognat Landré, C. L., *De besmettelijkheid der lepra arabum, bewezen door de geschiedenis dezer ziekte in Suriname* (Utrecht: J. L. Beijers, 1867), p. 4.
39 R. Edmond, *Leprosy and Empire: A Medical and Cultural History* (Cambridge: Cambridge University Press, 2006), pp. 44–60; M. Vollset, 'Globalizing Leprosy: A Transnational History of Production and Circulation of Medical Knowledge 1850–1930' (PhD thesis, University of Oslo, 2013).
40 Drognat Landré, *Besmettelijkheid der lepra*, pp. 72–5.
41 Drognat Landré, *Besmettelijkheid der lepra*, p. 77.
42 Drognat Landré, *Besmettelijkheid der lepra*, pp. 87–91.
43 Drognat Landré, *Besmettelijkheid der lepra*, p. 84.
44 Drognat Landré, *Besmettelijkheid der lepra*, p. 95.
45 Edmond, *Leprosy and Empire*, pp. 51–3; Michael Worboys, '"An Imperial Danger": Leprosy and Contagion, 1860–1900' (unpublished paper, University of Manchester, 2004).
46 G. Milroy (1871), quoted in Edmond, *Leprosy and Empire*, p. 63.
47 J. D. Hillis, *Leprosy in British Guiana: An Account of West Indian Leprosy* (London: J. A. Churchill, 1881), pp. 160–4, 182.
48 *Staatsblad voor Nederlandsch-Indië*, 1865, Bijblad, nr. 1715, 29-9-1865, see van Bergen, 'Vreeselijkste van alle kwalen', pp. 96–9.
49 SN: *De Tijd*, 23 January 1867.
50 G. A. Lindeboom, *Dutch Medical Biography: A Biographical Dictionary of Dutch physicians and surgeons 1475–1975* (Amsterdam: Rodopi, 1984), pp. 333–4.
51 B. Carsten, 'Over de verspreidingswijze van lepra', *Nederlands Tijdschrift voor Geneeskunde* 11 (1867), pp. 481–5.
52 Lindeboom, *Dutch Medical Biography*, pp. 1154–5.
53 F. J. van Leent, 'Review of Landré, *Besmettelijkheid*', *Geneeskundig Tijdschrift voor de Zeemagt* 7 (1869), pp. 63–75.
54 G. D. L. Huet, 'Een geval van lepra arabum. Lijkopening', *Nederlandsch Tijdschrift voor Geneeskunde* 12 (1868), pp. 113–20.

55 Van Leent, 'Review', p. 71.
56 C. L. Drognat Landré, *De la contagion seule cause de la propagation de la leprè* (Paris: Guillaume Baillière, 1869), pp. 58–9.
57 Appointment Uhlig: CR: *Koloniaal Verslag over 1856*, p. 24; honourable discharge: SN: *De West-Indiër* 11 October 1863. His report on Batavia is not preserved.
58 Van Leent, 'Review', pp. 73–4.
59 Drognat Landré, *De la contagion*, pp. 56.
60 Drognat Landré, *De la contagion*, pp. 50–5.
61 T. May, 'De lepra, haar voorkomen, verspreiding en bestrijding, in 't bijzonder in Suriname', I, *West-Indische Gids* 8 (1927), pp. 547–56, on pp. 550–1; H. Menke, S. Snelders and T. Pieters, 'Omgang met lepra in 'de West' in de negentiende eeuw. Tegendraadse maar betekenisvolle geluiden vanuit Suriname', *Studium* 2 (2009), pp. 65–77, on p. 75.
62 For biographical information on Vinkhuijzen: http://digitalgallery.nypl.org/nypldigital/explore/dgexplore.cfm?col_id=206. Accessed 13 May 2014.
63 H. J. Vinkhuijzen, *De melaatschheid, vooral met betrekking tot hare oorzaken en verhouding in de maatschappij* ('s-Gravenhage: De Gebroeders Van Cleef, 1868).
64 C. H. Sanders, *De melaatschheid* (Groningen: R. J. Schierbeek, 1867).
65 SG: 'Verslag aan de Tweede Kamer van het Geneeskundig Staatstoezigt in 1866–1867', Kamerstuk Tweede Kamer 1869–1870, nr. 39 ondernummer 1, 3.
66 SN: *De West-Indiër* 10 March 1867.
67 National Archive, The Hague, Handelingen van de Staten van Suriname (hereafter CE), 1867, p. 68.
68 CE 1868, p. 69.
69 SG: Tweede Kamer 1871–1872 stuk 54 ondernummer 9.
70 SG: Tweede Kamer 1871–1872 stuk 56 ondernummer 9.
71 SG: Tweede Kamer 1873–1874, Kamerstuk 66 ondernummer 9.
72 C. Landré, 'Sur la contagion de la leprè', in F. J. van Leent, A. A. G. Guye, de Perrot, and J. Zeeman (eds.), *Congrès international de medicine des colonies, Amsterdam, Septembre 1883* (Amsterdam: F. van Rossen, 1884), pp. 277–9.
73 C. Landré, *Over de oorzaken der verbreiding van de lepra. Een waarschuwend woord hoofdzakelijk gericht tot de bewoners van Suriname* (The Hague: Martinus Nijhoff, 1889).
74 R. A. Reddingius, 'Review of Landré, *Oorzaken*', *Nederlandsch Tijdschrift voor Geneeskunde* 33 (1889), p. 586.

6
Towards a modern colonial state: reorganizing leprosy care, 1890–1900

The death of Father Damien in the Kulawao leprosy settlement on the Hawaiian island of Molokai in 1889 spread fears of leprosy as an 'imperial danger' across the world. Once again, the international medical community was convinced of leprosy's contagiousness and considered the advisability of compulsory segregation. These developments occurred during a reorganization of leprosy care in Suriname in the 1890s. However, this reorganization had a dynamic of its own tied to the heritage of Surinamese confinement policies and the necessity for an accommodation between the dominant Christian religious groups in the colony (Protestants and Catholics) and with the colonial state. A symbiotic alliance for leprosy care had formed between the colonial state and the Catholics earlier in the nineteenth century. However, at the end of the nineteenth century, this alliance was renegotiated within the transforming landscape of Surinamese society to incorporate Protestants as well. The reorganization of leprosy care in the colony was intended to establish a better-organized leprosy asylum that addressed three issues. The first of these was that the asylum should be more accommodating to the citizens of a 'modern' colonial state than the dumping grounds of Voorzorg and Batavia had been. Second and third, the government should acquiesce to pleas from medical doctors for more humane treatment and manage the interests of religious groups who wanted to maintain or gain a foot in leprosy care. However, the new care steadfastly continued the traditions of contagionism, compulsory segregation, and racist prejudices that had characterized Surinamese leprosy politics since the eighteenth century, long before the international concerns of the 1890s.

The reorganization of leprosy care occurred within a context characteristic of the development of Dutch society – a phenomenon that Dutch political scientists and historians call 'verzuiling' ('pillarization'). The term refers to a process by which society is divided into various political and denominational groups, or so-called 'pillars' (Dutch: 'zuilen'). The everyday lives of members of each pillar are segregated from the lives of members of other pillars, even though they might live in close geographical proximity. Each pillar has its own schools, universities, social and cultural associations, and political parties. While social interaction between members of various pillars is limited, the top members of each pillar constitute an elite group that negotiate policies. This process of pillarization is regarded as characteristic of the modernization of Dutch society, starting in the later nineteenth century and coming to full fruition after the First World War.[1]

The pillarization of leprosy care in Suriname was an early instance of general pillarization across the Dutch empire that was triggered by the particularities of the Surinamese situation. Pillarization gave Catholics and Protestants the political and institutional space to take care of their own sufferers and incorporated the Moravians in addition to the Catholics in healthcare. In 1894, the Moravians were the largest religious group in Suriname with almost 25,000 members. Together with the smaller Calvinist 'Nederlandsch-Hervormde' ('Dutch Reformed') Church with almost 6,000 members, and 3,000 Lutherans, the Protestants numbered approximately 33,000, which represented half the Surinamese population. The Catholics were second but far behind with 11,000 members. By the late nineteenth century, the Hindus had become the third largest group with almost 8,000 followers.[2] However, it was not until after the Second World War that non-Christian groups gained a foothold in Surinamese politics. Leprosy sufferers who did not fit or wish to fit into the Catholic or Protestant asylums would be cared for in a government asylum.

The Peters report

The first signs of a renewed interest in leprosy care in Suriname appeared in 1891. In February, the Governor Jonkheer (esquire) Maurits de Savornin Lohman sent one of his military medical officers, O. A. Peters, on a journey to the British Caribbean colonies Guiana and Trinidad. Peters was to investigate the state of the art of leprosy care in those two countries.[3] For almost twenty years, leprosy had not been a major topic of public discussion or government attention in Suriname. De Savornin Lohman's motivation to send Peters on the trip is not exactly known. However, there was reason to reappraise leprosy care in Suriname. Three months before Peters received his assignment, the Catholic bishop Wilhelmus Wulfingh (1839–1906) had offered to take over all leprosy care from the colonial state and build a new asylum for the leprosy sufferers.[4] The timing of Wulfingh's offer was probably related to Father Damien's demise in Molokai. The death of the Catholic priest (and leprosy sufferer) in the Hawaiian leprosy asylum in 1889 made Damien into an international celebrity.[5] Media attention across the world stimulated renewed concern for leprosy, leading the British Empire to initiate investigations into leprosy in India and South Africa, and ultimately contributing to the organization of an International Leprosy Conference in Berlin in 1897. These British investigations rejected the hereditarian conclusions of the 1867 Royal College Report.[6] In response, fear entered colonial public health discussions as debate focused on the question of whether leprosy was contagious and if so, how great was the risk of contagion. The British investigatory activities explain why the Surinamese Governor sent his medical officer to British colonies close by.

Internationally, public and political unrest triggered by the Damien affair as presented in the international press and sensational books (such as publications by the Anglican priest Henry Press Wright in England) was combined with a reorientation of two interest groups – medical professionals and religious groups.[7] A renewed emphasis in the international medical community on the contagiousness of leprosy underscored the fears of the disease spreading throughout the European empires. The new (or returned) danger also led to new opportunities for religious missions as they took up the fight against leprosy and the care of leprosy sufferers. These groups were guided by various motives, including idealism, but also the possibility of strengthening their position in the colonies. At the 1897 Berlin conference, an international consensus was sought about the contagiousness of leprosy, the role of the Hansen bacillus in its aetiology, and necessary medical and public health policies.

In Suriname, no one felt the need to wait for this conference or its outcome. After all, a belief in contagionism and compulsory segregation had never been abandoned in the colony and renewed international fears only confirmed existing views. Peters' report took up where leprosy discussions in Suriname had left off twenty years before, and established a continuity of views and practices of modern colonial medicine with those of slave holder medicine.

The British medical officers in Guiana and Trinidad informed Peters that according to general scientific opinion, leprosy was contagious. This conclusion was not yet representative of international scientific discussions, but was completely in line with the Caribbean perspective on the disease. Therefore, Peters concluded that leprosy sufferers had to be segregated, even though knowledge of transmission mechanisms was still limited. However, Peters introduced a new approach to the practice of segregation. For Peters, segregation should not be done as it had been practised in the past, in a 'violent' way, but rather in an asylum with 'loving and careful nursing'. This meant not in Batavia, but instead in a new establishment not too far from Paramaribo, so that families could visit the sufferers. Peters also argued that sufferers should be allowed to marry, arguing that in Molokai only a small number of the children born there were afflicted with leprosy, a fact used to show that leprosy was therefore not a hereditary disease. When children born out of marriage in the new asylum turned two, they should be returned to their families outside the asylum and examined annually for signs of the disease. Peters also felt that the sufferers should live in one large building and not in many small houses as in Batavia, and that there should be a separate hospital and small gardens where the sufferers could work. The asylum would sell their products and give the sufferers half of the profits as pocket money. Peters recommended a new kind of asylum to replace the old asylum that was primarily a place of segregation and exclusion, not for medical care. Leprosy care should be humanized. The underlying idea was that the Surinamese, even if they were leprosy sufferers, now had to be treated as citizens of the modern colonial state.

Notwithstanding these shifts in ideas, Peters argued that leprosy sufferers still had to be forced to undergo this humanitarian treatment. The old racial prejudices born in the slave society were still present in the Peters report. According to the medical officer, the 'inboorlingen' ('natives') had little interest in material welfare, preferred to live together immorally in their slum dwellings, and did not fear contagion because they believed in the role of the treef. Therefore, Peters recommended that the edict of 1830 should remain the basis of leprosy policies so that the sufferers could be forced to live in a more healthy way.[8] Peters' report foreshadowed the new asylums founded in the next decade: Majella, Groot-Chatillon, and Bethesda. But questions remained: who should execute his ideas, and who would pay the bill?

Unrest in Batavia

Catholic bishop Wulfingh took the initiative for reorganization of leprosy care. In 1888, he had come from the Netherlands to Suriname as a Redemptorist missionary and a year later he became bishop and head of the Catholic mission in Suriname. The Redemptorists (the Congregatio Sanctissimi Redemptoristi, or in Dutch the Congregatie van de Allerheiligste Verlosser) were a congregation of Catholic monks focused on evangelization. In 1865, after the emancipation of the slaves, the Pope had placed the Dutch congregation of Redemptorists in charge of the mission in Suriname.[9] This meant that Peerke Donders (the resident priest in Batavia) had to go to Paramaribo in 1866 to become a member of the congregation and take his monk's vows. After almost a year in Paramaribo, Donders returned to Batavia in 1867.

One of the reasons for Wulfingh's interest in the findings of the Peters report was that Batavia was still a problematic battleground for both the Catholic missionaries and the colonial state. When Donders returned, it was obvious to all observers that something had to be done about the insufficient housing, clothing, and food rations in the asylum. The discharge of non-infected inmates after slave emancipation had not led to the expected financial savings.[10] By 1881, the buildings were in a sad state, and the situation worsened when the river dam broke and the asylum was flooded. Investments and repairs were overdue.[11] A member of the Colonial Estates visited the asylum and told his colleagues that where once one could have witnessed a pretty sight on arrival in Batavia, now all one saw was a jungle with decayed buildings, ruined drainage, and only four aged labourers to do repairs.[12] The ruined drainage was held responsible for the high incidence of malaria among the sufferers.¹³ In 1889, Herko Groot (who had been the district physician responsible for Batavia for twenty years) was rather satisfied with the health situation there, however he did comment that there were no beds or other utilities in the 'hospital' and that patients had to be treated in their dwellings.¹⁴

Another major problem was the unruliness and unrest among the leprosy sufferers and their families. Donders' first biography (or rather hagiography) relates how after his return to Batavia in 1867 he was once again confronted with the immoral and sensual lifestyle of the sufferers. Drunkenness and extramarital sexual relations were the order of the day. The Catholic priests even believed that healthy women settled themselves expressly in Batavia to have sexual intercourse with leprosy sufferers, whose lust had supposedly significantly increased as a result of their disease. As had been his task after Magneè's poisoning in 1849, Donders had to re-establish discipline in the asylum.¹⁵ Since he was now also responsible for missionary work on nearby plantations and among Amerindian tribes, he received assistance. Several Redemptorist priests replaced each other the following years in Batavia. Among them was R. P. Bakker, who was stationed in Batavia from 1883 until his death in 1890 and who was also infected with leprosy. His death, however, did not create the international stir that Damien's death had. Donders was stationed for two years in the Coronie district in 1883 and returned to Batavia in 1885, finally dying there in 1887 (but never having contracted leprosy).¹⁶

Even with the assistance of a second priest, Donders' task was not easy. Catholic sources complain about the unruliness of his flock, and identify a 'hard core' of ten to twenty people who refused to give up their immoral behaviour. Even the gift of an organ to the church in Batavia, considered a strong move to restore discipline among 'music-loving' Afro-Surinamese, did not seduce this hard core to attend church.¹⁷ However, there was not open resistance against all authority in Batavia. Instead, sufferers expressed what anthropologist James Scott has designated 'hidden transcripts' of anger and frustration, which was a discourse that had existed in Batavia alongside the public transcript of Christian care and obedience.¹⁸ Only rarely did this hidden transcript become public, such as during the Bishop's visit to Batavia in 1883. A deputation of sufferers addressed the Bishop, but since they spoke Sranan Tongo he was unable to understand them and had to call for

Donders to translate. It turned out that these 'ungrateful people' (as a Donders biographer designates them) asked that Donders be replaced. The priest was too old and they could not understand his sermons. It is interesting to note that the bishop responded to the request since Donders was sent to the Coronie district for two years following the Bishop's visit.[19]

Unruliness was expressed most dramatically by the Afro-Surinamese (the majority of sufferers) and far less frequently among sufferers with other backgrounds. This was possibly because the Afro-Surinamese had a long tradition of unrest reaching back into their slave past. According to Colonial Reports, the Afro-Surinamese always complained about their weekly provisions and were quite intolerant of each other. They insulted the director and fought among themselves.[20]

Other occasions during which resistance came out into the open occurred when healthy children were separated from afflicted parents and sent to their families in Paramaribo. In 1874, soldiers were sent to Batavia to enforce this removal and Donders had to act as an intermediary to ensure a peaceful conclusion of the procedure.[21] In 1892, one mother (perhaps not a leprosy sufferer) went so far as to go to Paramaribo to recover her son who had been placed in a Catholic orphanage one month earlier. Her mission succeeded and she took the child back with her to Batavia.[22] A letter from the director in 1897 makes it clear that at that time children were not sent to Paramaribo for examination of leprosy until they were eighteen years old, unless they were unruly or showed immoral behaviour.[23] If the children's examinations in Paramaribo were not done in a regular and systematic way, this would explain some of the unrest and frustration among the leprosy sufferers. Sometimes anger and frustration led to violence. In 1892, the district commissioner visited Batavia with eight soldiers to quell unrest.[24] In 1893, a leprosy sufferer who had beaten up the director and threatened the Catholic priest was jailed for two months.[25] Not all these incidents were publicly reported however; the Colonial Report of 1894 stated that in 1893, order had not been disturbed.[26]

Life in Batavia was more than frustration and violence, though. The German naval medical officer Peter Lens was stationed on the Dutch naval basis in Curacao and visited Suriname and Batavia in 1895.[27] His picture of the situation was more nuanced. He emphasized the Batavian inhabitants' practice of mutual aid, whether for sufferers or non-sufferers. Sufferers in whom the disease was less developed assisted those in a more advanced stage of the disease. The latter had creative solutions for their problems. Missing or useless fingers were replaced with pickaxes and other instruments tied to the wrist. Lens counted ninety-five sufferers, including six children, and forty-eight non-sufferers, including twenty children and women who lived together with infected husbands. Free love, as Lens called it, was widely practised in Suriname and mutual sexual intercourse between sufferers and non-sufferers occurred in Batavia. Fear of contagion seemed to be absent. Healthy boat crews and Amerindians freely mingled with the sufferers. Most of the sufferers left the premises at will to hunt and fish.

Lens did not notice religious problems. There were seven Moravian sufferers, but he did not witness Catholic pressure on them to convert. The problems he did notice were the neglected state of the sufferers' dwellings and their complaints about the food rations. People in the 'first category', that is, those who were not descended from slaves and the British Indians who were protected by the British Consul, received better rations than people in the 'second category', who were mainly Afro-Surinamese.[28] This would explain the latter's continuous complaints.

Batavia in the 1890s remained a rather anarchic place where administrators, priests, sufferers and their families continuously negotiated the situation, sufferers had wide freedom of movement, and strict segregation was an illusion. While the colonial state could tolerate this anarchic state since the main function of an asylum was segregation and exclusion far in the jungle, the new leprosy asylum as recommended by Peters' report demanded another approach.

The Catholics move

In the early 1820s, the Catholics seized their chance to assume the care of leprosy sufferers when the asylum moved from Voorzorg to Batavia. In 1890, they thought to seize it again and completely take over leprosy care. To Wulfingh, the idea of a new genuine Catholic asylum had several advantages. It would solve the problems in Batavia by simply eliminating them without losing the Catholic anchorage in leprosy care, and it would invite further financial contributions for his Catholic mission from the Netherlands. In 1889 and 1890, Wulfingh was back in the Netherlands for his ordination as bishop. He started a charity programme there to raise money for the mission.[29] Leprosy care was still a crucial element in Catholic propaganda. After returning to Suriname in December 1890, Wulfingh made the offer to the Governor to take over leprosy care and build a new asylum. In this way he would relieve the colonial state of the care for the leprosy sufferers. However, to do so, the Catholics would still need financial support from the colonial state.[30] Wulfingh's proposal was the exact reason for the Governor to send Peters on his fact-finding mission to the British Caribbean. The Bishop had reset the leprosy agenda in Suriname.

De Savornin Lohman's successor as governor of Suriname, Titus van Asch van Wijck, who took up office in 1891, was positive about the Catholic proposal. On his visit to Batavia, he found the situation there unbearable. There were now more leprosy sufferers living in Paramaribo than in Batavia, but how could he force sufferers to move against their will to an asylum if their objections to the place were so plainly justified? The situation in Batavia was deplorable and measures had to be taken. Why not leave it to the Catholics to solve the situation with little involvement from the colonial state?[31] He negotiated a contract with Wulfingh and asked for the Colonial Estates' approval in March 1892.

Van Asch van Wijck was a Protestant nobleman who had served as a member of the Dutch Parliament for the Calvinist Anti-Revolutionary Party. He was well aware of a major disadvantage of the new deal with the Catholics. In Batavia, the Catholics were allowed to be present and do religious work among the leprosy sufferers. They had no formal authority in the asylum; however, in the new asylum this would radically change. Would this also mean that the Catholic's religious control over the sufferers would be complete including Protestants, Hindus, and people who practised what was called a 'Chinese religion'? If so, this would make the new deal unacceptable to the Governor and the non-Catholic majority of the Surinamese population. Therefore, when the agreement was presented to the Colonial Estates, it contained an article that non-Catholic sufferers could be visited by their own religious attendants at agreed upon hours. If one of these sufferers was dying, he or she could ask for religious attendance from a representative of his own faith with permission of the attendant physician. With this article,

Van Asch van Wijck and Wulfingh hoped to eliminate Protestant objections to the agreement.[32]

Religious battleground

The two men were mistaken. Their agreement with the government led to a hue and cry in the Protestant community, both in Suriname and in the Netherlands. Once again leprosy care turned into a battleground for religious conflict. Immediately after the news of the agreement became public, Protestant churches joined forces and started to protest.[33] The Governor then asked Wulfingh to reconsider the agreement and to allow for the construction of a separate building on the premises of the new asylum where religious services of other denominations could be held. Wulfingh refused. He had already bought the ground for his new asylum, the plantation Groot-Chatillon on the bank of the Suriname River, three to four hours' journey by boat from Paramaribo, and he was only prepared to go so far to appease other faiths.[34]

When the agreement was discussed in the Colonial Estates in 1892, the colonial administrator of finances, the Baron Jacob Schimmelpenninck van der Ooije, aggressively defended the proposal against the Protestants. The Baron believed that religious freedom in the new asylum was ensured. There had never been problems with the Catholics, while he claimed that the Protestants, and especially the Calvinists and the Lutherans, had never shown much interest in the leprosy sufferers at Batavia. The defence was successful. In July, the Colonial Estates accepted the proposal under the condition that within a year construction of the new asylum would begin.[35] The Colonial Estates were persuaded in favour of the proposal by the fact that the contagious nature of leprosy was now considered proven by medical science, while Batavia was no longer in a condition to provide adequate housing and care to newly infected sufferers.[36]

Wulfingh's triumph was short-lived, however. Protestant resistance gained momentum. Led by the Calvinist minister of Paramaribo, H. H. Zaalberg, the Protestants threatened resistance if 'their Negroes' were forced to go to a Catholic asylum.[37] They appealed to the Queen Mother and Regent of the Netherlands, Queen Emma. By August, they had collected 1,200 signatories.[38] In the face of this clamour, the plan fell through. Nothing was built on the grounds of Groot-Chatillon. When in March 1893 the deadline was reached, the agreement was annulled. Wulfingh sold Groot-Chatillon to the government.[39]

Building a new consensus

In 1893, the Protestant churches also offered to take over leprosy care. To escape from this religious battle, and to reach a new consensus, the Governor set up a committee. The composition of this committee shows the process of pillarization at work. Both Catholics and Protestants were represented on the committee, as was the medical inspector. Schimmelpenninck van der Ooije took the chairmanship as representative of the colonial state. The committee's conclusions were presented in December 1893 and clearly sought a middle ground for all religious and medical divisions. The committee took its cue from British discussions following the publication of the report of the Leprosy Commission in British India in April 1893. This report had concluded that although leprosy was contagious, 'the amount of contagion which exists is so small that it may be disregarded'.[40] While the report took a stance against segregation, the *British Medical Journal* came to the opposite conclusion: if the possibility of contagion existed (however small), then 'men of the world' (i.e., colonial administrators) wished to segregate the sufferers. The Schimmelpenninck Committee decided to follow this last conclusion as well as an 1889 Trinidadian report stating that though the risk of infection was low, segregation (preferably under 'prison conditions') remained necessary to reduce the loci of infection.[41]

Dutch medical practitioners in general in the Caribbean in the 1890s held the idea that leprosy was a contagious disease with a rather low risk of transmission.[42] While attuned to contemporary British discussions, the opinions of these practitioners did not differ significantly from those that had been expressed by Drognat Landrè. Reference to the English medical literature only gave more authority to Surinamese medical views. Therefore, the colonial government in Suriname did not have to wait for the outcomes of the 1897 international Berlin conference, a conference where they were not even represented (although there were representatives of the medical officers in the Dutch East Indies in the Dutch delegation in Berlin).[43] That the debate on the roles of infection and inheritance had not yet reached a consensus in the Netherlands at the time was not of practical importance to Suriname. The Rotterdam dermatologist Thomas Broes van Dort, who was present at the Berlin conference, published an overview of the leprosy situation in the Dutch empire in a series of articles in the *Dutch Journal of Medicine* in 1897. Broes van Dort had not come to a definitive conclusion and even pointed to the possible role of a lack of fresh air in the aetiology of leprosy on some of the Antillean islands. Van Dort's medical informants in Suriname, among whom were Peters and medical inspector Salomons, were all convinced of the role of contagion, though they also believed in the role of predisposition as a risk factor.[44]

The Schimmelpenninck Committee advocated segregating sufferers, although it emphasized that the risk of contagion was not high and that a general medical consensus on transmission was lacking. Still, there were around 1,500 sufferers in Suriname according to the committee. As in the discussions in 1850, when the same number had come up, it was unclear where this number had come from. Salomons thought the number was unreliable, while Peters estimated a total of 500 sufferers.[45] Nonetheless, the high number of sufferers was one argument for the Schimmelpenninck Committee to recommend a new leprosy asylum. Unlike Batavia, discipline, hygiene, and care were to be central concerns of this new asylum. This was to be achieved through the provision of common sleeping rooms for ten to twelve people, and if people could pay for their treatment, they could have their own houses. Unmarried sufferers would live separated from the married ones and newly born children would be removed from the asylum as soon as possible. Capable people should perform labour. The director would also be the physician in charge of the hospital. So far, this was in essence what the Peters report stated, but a religious compromise was added. There would be two departments, one for Catholics and one for Protestants. The Schimmelpenninck Committee also felt that in food preparation, religious customs should be observed, including the sufferers' notions about treef.[46] To settle the dispute between the religious groups, and to reach a new consensus, the Schimmelpenninck Committee advocated a pillarization within a new leprosy asylum. However, for the churches this still went too far. Pillarization did take place, but not in one location or in one asylum.

Private asylums

In the meantime, Wulfingh did not relinquish his dream of a separate Catholic asylum. If he could not take charge of all leprosy care, he could at least build his own private asylum for Catholic sufferers. To do so, he forged a new alliance with a group of Catholic nuns from the southern Dutch province of Brabant, the Zusters van Liefde (Sisters of Love).[47] The nuns would provide the nurses for the new Catholic asylum. In September 1894, the first group of six sisters arrived in Paramaribo. Wulfingh also bought new grounds for the asylum next to the military hospital in Paramaribo. To administer the asylum, a Gerardus Majella Stichting ('Gerardus Majella Foundation') named after an eighteenth-century Catholic saint was founded, and in October 1895 the first four Catholic leprosy sufferers were admitted to the asylum. By 1900, there were almost sixty sufferers.[48]

The colonial government did not have problems with a separate asylum for the Catholic leprosy sufferers. Pillarization in separate asylums had the added advantage that it was cheaper for the government. Building a new asylum would cost almost 350,000 guilders, while its operation would cost almost 125,000 guilders each year if there were 200 patients. Private asylums could reduce the costs for the colonial state even when they received financial support of 4,000 guilders each year, plus 100 guilders extra for each patient.[49] The only problem with the new Catholic asylum was its location on the outskirts of Paramaribo. This ran counter to the tendency of the previous one and a half centuries to build an asylum as far away from the inhabited world as possible, and it increased the fears of the Europeans in the colony that they might have direct contact with leprosy sufferers. Salomons' Medical Committee allayed these fears. Salomons was convinced of the value of the Catholic contribution since he thought the Catholics were fearless of the risk of infection, and he believed that they would ensure religious freedom.[50] On 18 July 1895, he signed a contract with Wulfingh to place leprosy care in Suriname on a new basis at last. The contract was a remarkable example of public–private partnership going far beyond the cooperation of the colonial state and the Roman Catholic Church in Batavia. Wulfingh agreed to take in any leprosy sufferer in Majella sent there by the government. He also agreed that non-Catholic patients would not be addressed on religious issues. In return, he would receive 100 guilders per year for each patient.[51]

In July 1896, the Medical Committee inspected Majella and concluded that the situation was satisfactory and that there was no danger of leprosy spreading from the asylum to the town.[52] Majella was now on track. In order to strengthen the image of the mission in the Netherlands and the alliance with the Sisters of Love, Donders was put forward as a possible candidate for beatification and ultimately sainthood. The first session of the church process necessary for the beatification started in November 1900.[53] That Donders had been born in Tilburg (where the headquarters of the Sisters of Love was based) played an important part in the decision to put forward Donders as a Dutch version of Damien, rather than Bakker who had died from leprosy in Batavia in 1890, or other Catholic priests and nuns who later suffered from leprosy.[54] However, it would take until 1982 before Donders was beatified, and the church process to elevate him into sainthood is still going on at the moment of this writing in the early twenty-first century.

While the Catholics finally built their own asylum, preparations were going on for a new state asylum on the land of Groot-Chatillon bought from the Catholics. The Protestants were no longer content to be outmanoeuvred by the Catholics as they had been in Batavia, and they wished to build a church for Protestant religious services in the new state asylum. Zaalberg, the minister of the Calvinist church in Paramaribo, who had played a central role in the protest against the original deal of the Governor with the Catholics, became one of the founders of a Protestant Society for the Care of Leprosy Sufferers in the Colony of Suriname. In 1897, Zaalberg went to the Netherlands to collect funds for the new church in Groot-Chatillon among the Protestant communities. On his trip, he collected 7,000 guilders and a second funding campaign the next year raised 18,000. The response was such that it was decided not to just to build a church, but rather a whole Protestant asylum next to Groot-Chatillon, to be called Bethesda and run by the Moravians while the Protestant Society organized the funding. Bethesda received recognition from the government, and on 19 July 1899 the first leprosy sufferer moved into one of the two small houses then existing in Bethesda accompanied by one nurse. The sufferers were welcomed by the Moravian religious teacher and his wife. By 1900, the number of patients had already grown to ten.[55]

The new state asylum

By 1900, the Catholics and the Protestants had their own asylums. Leprosy sufferers who were admitted there were new patients. These asylums started fresh. In principle, all leprosy sufferers who were compulsorily segregated by the leprosy committee would be sent to the state asylum Groot-Chatillon, though they could ask for admittance to Majella or Bethesda. Significantly, the Catholic leprosy sufferers in Batavia, who formed the large majority of the people living there, were transferred to the new state asylum, not to Majella. Sending the sufferers of Batavia to Groot-Chatillon was clearly advantageous to the Roman Catholic Church. The nominally Catholic leprosy sufferers in Batavia, with their long tradition of unrest and hidden transcripts of resistance would not disturb the new arrangements of Majella.

Leprosy sufferers in Batavia were not too keen on being transported to Groot-Chatillon, just as their predecessors in Voorzorg had resented their transportation to Batavia. In 1897, the transfer was scheduled to take place. Around Christmas and New Year's Eve 1896, there were fears among the Batavian administrators that the sufferers would disturb the festivities. A military officer was stationed in the asylum, but there were no disturbances.[56] Nevertheless, as in 1824, the government enlisted the help of the Catholics to ensure a tranquil transportation to Groot-Chatillon. With the exception of one British Indian Hindu and a few Afro-Surinamese, all the sufferers were Catholic at the time and there had been much unrest among them over the past year.[57] Wulfingh responded to the call of the government. On 27 September, Wulfingh, two priests, and one monk accompanied eighty-one sufferers on the boat from Batavia to Groot-Chatillon. This was the end of Batavia. On 4 October, the government had the buildings of the leprosy sufferers burned down. Three days later, unknown arsonists burned down the church and the vicarage as well, though there were still police stationed on the terrain. Only the doctor's house was untouched by the fire.[58] The last remains of leprosy care from the age of slavery were finally destroyed by violence.

Conclusion

The result of the reorganization of leprosy care in the 1890s was not a response to international discussions and panic about the disease as an 'imperial danger'. Neither was it a response to changing medical views on the transmission of the disease or the return of contagionism. Rather, developments in Suriname showed once again a distinct dynamic. Several developments combined to change the fate of the leprosy sufferers in Suriname: dissatisfaction with the failure of Batavia, the need for a well-organized establishment fitting for patients who were citizens of the modern Suriname, increased influence of medical doctors pleading for more humane treatment, and religious group interests in holding or gaining a firm foot in leprosy care. The reorganization took place in the context of an accommodation between dominant religious groups. Those denominations that did not have followers in the Dutch ruling elite (Winti practitioners, British Indian Hindus, or Javanese Muslims) were left out. As in the age of slavery, the organization of leprosy care mirrored the power structure and racist viewpoints of Surinamese society.

Notes

1 A. Lijphart, *The Politics of Accomodation: Pluralism and Democracy in the Netherlands* (Berkeley, CA: University of California Press, 1975); J. H. C. Blom and J. Talsma, *De verzuiling voorbij. Godsdienst, stand en natie in de lange negentiende eeuw* (Amsterdam: Het Spinhuis, 2000).
2 J. Vernooij, *Barmhartigheid een levensprogram. Zusters van liefde van Tilburg 100 jaar in Suriname* (Paramaribo, n.p., 1994), pp. 13–14.
3 Archive Bisdom of Paramaribo (hereafter BP): T5.
4 T. Lens, 'Lepra in Suriname', *Elsevier's Geïllustreerd Maandschrift* 10 (1895), pp. 521–52, on pp. 544–5.
5 For a biography of Damien, see G. Davis, *Holy Man: Father Damien of Molokai* (Honolulu, HI: University of Hawai'i Press, 1973).
6 For British reactions on the death of Damien, see R. Edmond, *Leprosy and Empire: A Medical and Cultural History* (Cambridge: Cambridge University Press, 2006), pp. 92–9, 158–60.
7 Books Press Wright: *Leprosy and Segregation*, published in 1885, and *Leprosy: An Imperial Danger*, published in 1889. See J. Robertson, 'Leprosy and the elusive *M. Leprae*: Colonial and imperial medical exchanges in the nineteenth century', *Manguinhos*, 10; suppl.1 (2003), pp. 13–40, on pp. 27–31; Edmond, *Leprosy and Empire*, pp. 84–5, 92.
8 Report Peters: BP: T5.

9 On the Redemptorists in Suriname: F. Schweigman, *Twee Missionarissen onder de Melaatschen en Indianen van Suriname* (Roermond: J. J. Remen, 1894), pp. 141–2.
10 National Archive, The Hague, Handelingen en Bijlagen van de (Koloniale) Staten van Suriname (hereafter CE), 1866/67, 18667/68, 1868/69.
11 CE 1882/1883, p. 70.
12 CE: Cateau van Roosevelt in the Colonial States session of 13 July 1882, CE 1882/1883, pp. 71–2.
13 Colonial Report (hereafter CR) 1897, p. 19.
14 National Archive, Paramaribo, archive Gouvernementssecretarie Suriname 1829–1954 (hereafter GS), 1224, submitted report for the Colonial Report of 1888. On Groot: BP: T1, 'Kronijken', entry 13 May 1892.
15 Schweigman, *Twee Missionarissen*, pp. 142–4; report Donders 10 October 1875, in BP: T1, 'Kronijken'; J. A. F. Kronenburg, *De eerbiedw. dienaar Gods Petrus Donders C.ss.R. Nieuwe levensbeschrijving* (Tilburg: W. Bergmans, 1925), p. 209.
16 List of priests in Batavia in Kronenburg, *Eerbiedw. dienaar Gods*, p. 170.
17 Schweigman, *Twee Missionarissen*, pp. 144–5.
18 J. C. Scott, *Domination and the Art of Resistance: Hidden Transcripts* (New Haven, CT: Yale University Press, 1990).
19 Schweigman, *Twee Missionarissen*, pp. 151; Kronenburg, *Eerbiedw. dienaar Gods*, p. 300.
20 CR 1882; submitted report for the Colonial Report of 1885, GS 1222; submitted report for the Colonial Report of 1888, GS 1224.
21 Kronenburg, *Eerbiedw. dienaar Gods*, p. 204.
22 BP: T1, 'Kronijken', entry 9 April 1892.
23 National Archive Paramaribo, archive Districtscommissariaat Saramacca 1897–1955 (hereafter DS), 413, book of letters 1897.
24 BP: T1, 'Kronijken', entry 22 August 1892.
25 BP: T1, 'Kronijken', entry 19 February 1893.
26 CR 1894, p. 18.
27 G. A. Lindeboom, *Dutch Medical Biography: A Biographical Dictionary of Dutch physicians and surgeons 1475–1975* (Amsterdam: Rodopi, 1984), p. 1173.
28 Lens, 'Lepra in Suriname', pp. 531–44.
29 *Vijftig jaren Hofbauer liefdewerk 1890–1940* (Rotterdam: Secretariaat der Surinaamsche Missie, 1940).
30 Lens, 'Lepra in Suriname', pp. 544–5.
31 CE 1891/1892, pp. 325–8.

32 CE 1891/1892, pp. 321–3. For the whole discussion, see J. Vernooij, *De rooms-katholieke gemeente in Suriname. Handboek van de geschiedenis van de Rooms-Katholieke Kerk in Suriname* (Paramaribo: Leo Victor, 1998), pp. 71–5. See also Lens, 'Lepra in Suriname', pp. 544–52.
33 Vernooij, *Rooms-katholieke gemeente*; Lens, 'Lepra in Suriname'.
34 Lens, 'Lepra in Suriname', p. 547.
35 CE 1892/1893, pp. 9–23.
36 CE 1892/1893, p. 18.
37 Zaalberg in *Stemmen voor Waarheid en Vrede*, 13 July 1892, BP T 19.
38 Lens, 'Lepra in Suriname', pp. 550–1.
39 A. C. Schalken, 'Historische gids bestaande uit chronologische lijst naamlijsten varia registers. 300 jaar R.K.-gemeente in Suriname 1683–1983' (Paramaribo, 1985), pp. 49–50.
40 Cit. in Edmond, *Leprosy and Empire*, p. 101.
41 Cit. in Edmond, *Leprosy and Empire*, p. 101.
42 Lens, 'Lepra in Suriname', p. 522.
43 *Mittheilungen und Verhandlungen der internationalen wissenschaftlichen Lepra-Conferenz zu Berlin*, 3 vols. (Berlin: August Hirschwald, 1897); S. Pandya, 'The first international leprosy conference, Berlin 1897', *Manguinhos* 10, suppl. 1 (2003), pp. 161–77; Edmond, *Leprosy and Empire*, pp. 103–7.
44 T. Broes van Dort, 'Een en ander over de lepra in Nederland en zijne koloniën', *Nederlandsch Tijdschrift voor Geneeskunde* 41 (1897), pp. 292–6, 384–91, 407–21, 650–1. On the opinions of doctors in Suriname: pp. 418–19. On the Berlin conference: T. Broes van Dort, 'De internationale lepra-conferentie te Berlijn (11–16 Oct. 1897)', *Nederlandsch Tijdschrift voor Geneeskunde* 41 (1897), pp. 747–71, 810–15, 893–7, 937–42, 978–84.
45 Broes van Dort, 'Een en ander', pp. 418–19.
46 BP: T30.
47 J. Brouwers, *Na de drie begijnen ging het verder. Gechiedenis van de Congregratie van de Zusters van Liefde van Onze Lieve Vrouw, Moeder van Barmhartigheid* ('s-Hertogenbosch: Congregratie Zusters van Liefde, 2000), pp. 98–120.
48 Schalken, 'Historische gids', pp. 51–3; Vernooij, 'Barmhartigheid'; number of patients BP T178.
49 Vernooij, *Rooms-katholieke gemeente*, pp. 74–5.
50 CE 1892/93, pp. 20–1; J. M. Plante Fébure, *West-Indië in het parlement 1897–1917, Bijdrage tot Nederland's koloniaal-politieke geschiedenis* ('s-Gravenhage, Martinus Nijhoff, 1918), pp. 52–3.
51 A copy of the contract in BP: T34.

52 BP: T149; Vernooij, *Rooms-Katholieke gemeente*, p. 75.
53 Schalken, 'Historische gids', pp. 56–7.
54 Email communications Joop Vernooij to the author, June 2014, March 2015.
55 Utrecht Historical Archive, Utrecht, Archive Zeister Zendingsgenootschap, inv. nr. 48-1 1230; *Verslag der herdenking van het 25-jarig bestaan van "Bethesda" op 15 Mei 1924* (Paramaribo: Bethesda, 1924), pp. 7–9; J. Postma, 'De leproserie Bethesda tussen 1897 en 1928', *OSO* 22 (2003), pp. 69–81, on pp. 69–70.
56 DS: 413, letter of the doctor in Batavia.
57 BP: T42, letter 7 August 1897.
58 DS: 413; BP: T 45b; Schalken, 'Historische gids', pp. 54–5.

7
Developing modern leprosy politics, 1900–1950

By 1900, the leprosy asylum system in Suriname was reorganized and adapted to both the modern humanitarian and medical demands of the time and the new accommodation between the colonial state and dominant religious groups. Attention could now turn to another essential and long-postponed reform: the legal framework and execution of the policy of compulsory segregation. Since the international leprosy conference in Berlin in 1897, the international medical and colonial world had been catching up with the supposed need for sufferer segregation in the fight against leprosy, however in Suriname this policy in and of itself had never really been in question. The problem here was not whether or not to organize segregation policies, but rather how to update them. A slow reframing of leprosy policies in Suriname took place, leading to the new leprosy edict of 1929 that replaced the edict of 1830. The new edict inaugurated a renewed era of increased detection and segregation of sufferers. By the 1940s, the colonial state claimed that leprosy was finally under control. However, this claim could be disputed.

Modern leprosy politics in the twentieth century were a Janus head. On the one side, the politics were based on the latest developments and fashion in medical views on leprosy: sufferers should be treated as patients, not as criminals; medical treatment in asylums and in outpatient clinics should be encouraged; and a humane organization of life in the asylums should be promoted. However, unlike in other colonies, such as British India or the Dutch East Indies, the idea of compulsory segregation was never abandoned. Sufferers with non-European backgrounds, especially the Afro-Surinamese and the British Indians, were still stigmatized and seen as unwilling or unable to cooperate. They had to be forced into segregation. On the other side of the Janus head, policies of surveillance, detection, and compulsory segregation were therefore intensified.

Modern leprosy politics were developed in the context of a modernizing colonial state and in an era of what has been labelled 'authoritarian high modernism'.[1] The twentieth century was the high age of modernity, change, and movement. This was expressed in a desire to order society rationally on the basis of, and with a belief in, the progress of scientific and technological knowledge, and by the exercise of authoritarian state power. The conditions of authoritarian rule in the colonies could enhance the possibilities and desirables of high modernist politics. Modern leprosy politics of the 1930s and 1940s coincided with the political high tide of a culture of domination and authoritarian colonial rule in Suriname.

The architects and executors of public health and leprosy politics in Suriname were not necessarily conscious 'modernists'. However, all the characteristics of authoritarian high modernism as developed in the writings of anthropologist James C. Scott are clearly recognizable in their activities. The public health authorities believed in a rational ordering of public health based on the progress of scientific and medical knowledge. They believed in the exercise of state power, and in politics of compulsory segregation, in order to achieve their goal. They worked to engineer socially a sub-population in the colony – the leprosy sufferers – who were seen as a potential threat to public health. In addition, they examined and redesigned inherited human habits and practices that were not based on scientific reasoning, such as those related to a belief in the treef. The public health professionals' authoritarian high modernism encountered contestation and resistance; thus the invasion and surveillance of the population's private sphere was never complete. Nevertheless, because colonial medicine could exercise the power of the colonial state, it was able to go far in developing new policies.[2]

Compulsory segregation, introduced in 1790 and codified in 1830, was emphasized again after 1929. Segregation was part of the colonial state and Western medicine's interventions to rule the bodies of the colonial subjects in a process that David Arnold described as the 'colonization of the body'.[3] In Suriname, this colonization was determined not only by the state of Western scientific and medical knowledge, but also, and inescapably, by racial conflict and labour management problems. Leprosy remained primarily a disease connected to race and labour, and especially connected to Afro-Surinamese descendants of the slaves and British Indian indentured migrants. This led to resentment of domination by the colonial rulers including medical doctors. As Arnold emphasizes, contestation was a central element in the dialectics of power and knowledge in colonial society. Thus Western colonial medicine had to address the limits of its practical possibilities. Although many sufferers and their families welcomed the medical and social benefits of leprosy politics, an 'infrapolitics' of non-cooperation and non-compliance in Suriname continued to exist, as it had in the age of slavery.

Reformulating the leprosy edict: failed attempts

By 1900, three new leprosy asylums were in operation. In response, the leprosy committee renewed its activities with more enthusiasm. Instead of having to condemn sufferers to involuntary segregation in a decrepit asylum that was two days' travel from Paramaribo in the jungle, it now had the option of segregating them in asylums relatively close or even next to town where more modern medical facilities and care were available.

This led to a slow increase in the number of sufferers in the asylums, which continued until 1925. From 1926, the increase in sufferers became more pronounced (see Table 3 and Figure 3), and by the end of the 1920s the absolute figures reached those of the heyday of the 'Great Confinement' policies and rose above 400. By then the population had grown to almost 150,000, so the relative figures were still much lower than in the nineteenth century.

Around 1900, policy makers and medical practitioners felt that leprosy detection was failing. In 1896, medical inspector Salomons told the Colonial Estates that only leprosy sufferers who were seen in public streets ended up in an asylum. Moreover, racial and sexual fears about leprosy sufferers survived. By 1900, Dutch colonial doctors assumed that British Indian and Chinese migrants contributed to the spread of the leprosy bacillus. The Chinese were quantitatively an insignificant minority among the sufferers; however, the opposite was true for the British Indians who were seen to be particularly disposed

Table 3 Numbers of patients in the asylums, 1896–1949

Year (per 31 December)	Groot-Chatillon	Majella	Bethesda	Total
1896		22		
1897	98			
1898	118	41		
1899	136	41		
1900	146	58	10	214
1901	139	58	10	207
1902	149	53	15	217
1903	146	62	20	228
1904	141	68	18	227
1905	116	70	23	208
1906	127	82	31	240
1907	123	83	33	239
1908	123	85	32	240
1909	104	104	38	246
1910	133	128	42	303
1911	108	115	47	268
1912	106	115	51	274
1913	106	105	56	267
1914	116	112	52	280
1915	135	103	45	283
1916	130	113	52	295
1917	136	112	53	301
1918	141	117	53	311
1919	146	114	52	312
1920	145	123	56	324
1921	135	123	55	313
1922	136	131	57	324
1923	144	119	61	324
1924	151	117	61	329
1925	154	110	57	321
1926	167	117	49	355
1927	161	157	52	370
1928	166	176	66	408

Table 3 (*cont.*)

Year (per 31 December)	Groot-Chatillon	Majella	Bethesda	Total
1929	165	168	74	407
1930	158	194	87	439
1931	194	201	87	482
1932	215	208	104	527
1933	221	185	110	516
1934	223	191	130	544
1935	234	203	136	573
1936	238	207	150	595
1937	233	175	159	557
1938	236	180	172	588
1939	240	180	167	587
1940	249	186	166	601
1941	269	224	182	675
1942	278	229	181	688
1943	269	234	182	685
1944	289	233	178	700
1945	299	234	176	709
1946	296	230	180	706
1947	299	242	185	726
1948	298	218	183	695
1949	291	234	189	714

Source: Colonial Reports; reports of the Public Health Service: GS 1258, 1262, 1264, BP T; Hallewas, 'Gezondheidszorg', App. XI (voor na 1918); Majella 1896–1898: BP T178.

to leprosy. In 1902, sixty-four of the 149 sufferers in the government asylum Groot-Chatillon were British Indians.[4] Similar to their perceptions of the Afro-Surinamese, the Dutch associated leprosy with British Indians' unhygienic customs, cramped living conditions, and unhealthy diet.[5] However, racialized fears of leprosy continued to focus on the Afro-Surinamese who constituted the majority of leprosy sufferers. According to Salomons, the necessary boundaries between

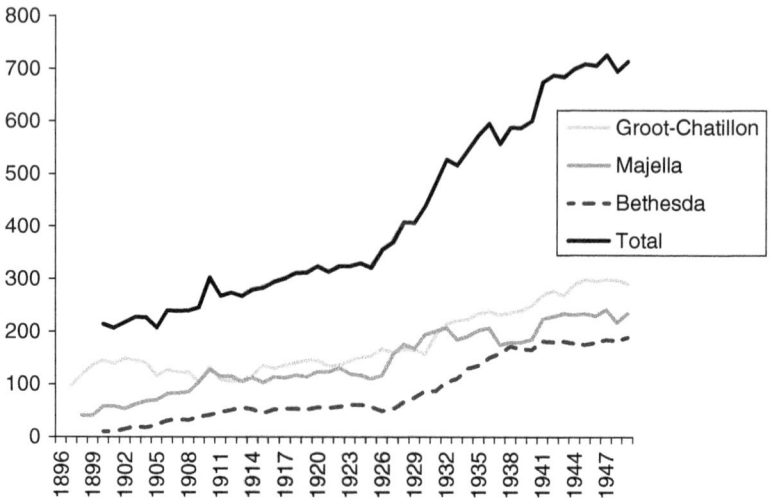

Figure 3 Numbers of patients in the asylums, 1896–1949

sufferers and non-sufferers were nearly non-existent among the Afro-Surinamese since sufferers freely participated in economic activities, such as cattle breeding, milk production, pastry making, ice and lemonade making, and tailoring. The Afro-Surinamese's potentially contaminated products were distributed by healthy marketers and street hawkers. In Salomons' stigmatizing view of the Afro-Surinamese, sexual elements were also present. He believed that female leprosy sufferers worked as prostitutes in their own homes and that male leprosy sufferers lived with healthy women.[6]

When in 1902 Cornelis Lely came to Suriname as Governor, he found that the general opinion was that the prevalence of leprosy was at the very least not decreasing and that a large number of undetected leprosy sufferers were living outside of the asylum.[7] In response, Lely and other Surinamese policy makers' attention turned to the basis of leprosy politics – the 1830 leprosy edict. Although still in operation, this edict was considered outmoded by the beginning of the twentieth century. It was framed in the age of slavery and based on the unfree status of the vast majority of segregated sufferers, whereas now all sufferers had civil rights. Moreover, the British Consul held watch over the legal interests of the British Indians. The 1830 edict made provisions for managing elephantiasis sufferers, although by now this disease was reconstructed as filariasis with another aetiology. Mosquitoes were regarded as filariasis vectors that primarily affected the lower classes because of poor hygienic conditions. To manage filariasis, improvement in these conditions among the population of Paramaribo were deemed to be the most effective measures.[8] Furthermore, the emphasis in leprosy debates had changed and there was a call for a more medical rather than a judicial or law enforcement approach to the problem of segregation. Finally, there were now three leprosy asylums, of which two were run by private organizations and not by the state and this demanded legislative adjustment. A new legal basis for leprosy politics was necessary.

Lely put his civil servants to work. In October 1904, he presented a new Lepraverordening or leprosy regulation to the Colonial Estates. It opened with the explicit statement that all leprosy sufferers both inside and outside of the asylums would be under the supervision of the colonial government. In a separate memorandum, the Norwegian system of segregation was held up as the global example of best practice since it had led to a decrease in leprosy. However, one all-important modification was deemed as necessary in Suriname since voluntary segregation was not feasible, private dwellings could not be isolated and the population was supposedly unafraid of the disease. The new edict therefore combined the creation of improved medical provisions in the asylums with an intensified regime of detection and surveillance. Under the new regulation, medical practitioners, district wardens, and school teachers were legally obligated to report cases of leprosy to the government. The police were ordered to arrest people suspected of having leprosy, including people with suspect signs noticed in the streets. The police officers continued to receive bounties for arrested sufferers. Infringements of the regulations could lead to prison sentences of up to six months and fines of up to 1,000 guilders. Once a leprosy committee of five medical doctors, chaired by the medical inspector, had decided on the necessity of admission to an asylum, the dwellings of those diagnosed with leprosy were to be disinfected or destroyed. While medical provisions in the asylums were improved, isolation was intensified as well. Asylums had to be situated in places with enough space for exercise and healthy air, but where the sufferers would not have contact with the outside world. The drainage of the asylums should not endanger public health, that is, not carry leprosy poisons into the inhabited world outside. Objects used by the sufferers, including letters, should not leave the asylums unless they were disinfected. Within the asylum, men and women should live segregated from each other, except when they were married; children born of these marriages had to be separated from their parents and moved out of the asylum as soon as possible. There was one exception to the rule of complete isolation. The Majella asylum, though situated close to town, was allowed to continue at its location for practical reasons. The Governor explained to the Colonial Estates that the location of the Majella asylum was ill-chosen, but the existence of the asylum was beneficial.[9]

The draft of the new leprosy edict was in the tradition of the leprosy regulations of 1790 and 1830, with the exception of the inclusion of private asylums. In the final analysis, the draft confirmed the accommodation between the colonial state and the various religious groups in Suriname that had led to the founding of the private asylums. With the new draft edict, Lely modernized the legal framework of the existing practice of leprosy politics.

However, the draft leprosy edict of 1904 was never put to a vote in the Colonial Estates, although the required report on the draft by a committee of members of the Estates was positive. The committee only wished for stricter surveillance and a thorough investigation of whether or not the location of Majella was really safe for the inhabitants of Paramaribo. The committee was convinced of the contagious nature of leprosy, and (with the exception of one committee member) that leprosy was on the increase in Suriname. They wished for a more exact investigation of this increase.[10] This would paradoxically lead to the draft edict of 1904 ending up in a desk drawer.

Lely's successor Alexander Idenburg agreed that more detailed evidence of the exact number of leprosy sufferers was needed. Idenburg had been Minister of Colonies before he came to Suriname in 1905, and was a propagator of the so-called 'ethical policy' in the Dutch East Indian colonies. The ethical policy aimed to create social policies and Western-style legislation for indigenous peoples. Idenburg had supported the Protestant missions' participation in leprosy care in the Dutch East Indies out of both ideological motives (he was a member of the Protestant Anti-Revolutionary Party in the Netherlands) and financial ones. In doing so, he partly transferred the leprosy policies in Suriname and the religious accommodation of the 1890s from the West to the East Indies. When taking over the governorship of Suriname, Idenburg wished to know more details about the prevalence of leprosy. In May 1906, he ordered a committee to determine the exact number of leprosy sufferers in Suriname. This investigation prompted a general inspection of all dwellings in Suriname with the compulsory cooperation of the inhabitants as provided for in the edict of 1830, but never executed since the end of slavery. The proclaimed aim of the survey was now not to detect and segregate sufferers, but rather to estimate their number. The survey committee did not include medical practitioners. Doctors were legally obligated to give full disclosure of their patients to the investigators. The detection of sufferers would not have consequences for the sufferers, however, which was a promise that was meant to ensure the population's cooperation with the survey.[11] This cooperation was considered problematic from the start. One member of the Colonial Estates thought that because of the stigma of leprosy, family members would hesitate to report their kin, although the member added that this would not be so much of a problem among British Indians and the 'lower classes', who presumably had less social shame than the middle and higher classes.[12]

In August 1909, the results of the survey were confidentially sent to the Colonial Estates. The survey discovered 239 sufferers segregated in the asylums, and approximately 400 sufferers living outside the asylums. This was a lower number than expected.[13] Based on the survey in 1909, the Minister of Colonies reported that 'only' approximately 0.75 per cent of the population was suffering from leprosy.[14] However, there was no certainty that sufferers had been detected who were unwilling to cooperate with the survey. The at-first-sight comforting conclusions of the leprosy survey could be interpreted in two diametrically opposite ways. If one accepted the results of the survey, the situation was not as serious and threatening as expected and the urgency of the new leprosy edict seemed to disappear from the project. The survey committee concluded in 1909 that the policy of compulsory segregation should not be executed more harshly, but rather that sufferers should be given more opportunities to segregate themselves voluntarily in their own homes. This had financial benefits, since less money would have to be provided to the asylums.[15] Given that additional money had been requested for Groot-Chatillon in December 1908, this idea was attractive to the colonial government.[16] However, on the other hand, fears about leprosy were not entirely assuaged in the colonial government because of the uncertainty whether all sufferers had been detected. There was no consensus and the discussion of the edict in the Colonial Estates was repeatedly postponed at the request of the Governor.[17] In 1916, the Minister of Colonies told the Dutch Parliament that the Colonial Estates were still working on improved regulations for leprosy sufferers.[18] A new version of the draft edict was presented to the Colonial Estates in June 1919. The provision that police officers receive a bounty for arrested leprosy sufferers was deleted, but otherwise the modifications were only minor.[19] However, since there was no consensus about whether leprosy was actually on the increase, this draft was also never put to vote and it took another seven years before consensus was reached and a new draft was sent to the Colonial Estates.

Continuing fears

Notwithstanding the reassuring words of the Minister of Colonies and the leprosy survey committee, the fear of leprosy among colonial officials did not subside, and the search for leprosy sufferers continued after 1909. The instrument of bounties for police officers detecting leprosy sufferers was increasingly employed. In 1911, three-quarters of the total of budgeted bounties for police officers identifying leprosy sufferers was spent.[20] The 1918 budget again provided for bounties for police officers.[21] Fears among colonial officials were stimulated when leprosy sufferers were discovered in Paramaribo. In November 1917, the Military Hospital was evacuated when the manager of hospital catering was diagnosed with leprosy. The building was disinfected and repainted both inside and out.[22] The 1916 budget contained 900 guilders for a post of police officers guarding suspected leprosy sufferers in the Military Hospital, a possible source of contagion for the hospital personnel.[23] In 1919, the adjunct medical inspector of Suriname warned that hundreds of leprosy sufferers were hiding in their houses and were focal points of contagion and hence a danger to 'all races' and even to foreign countries.[24] Once again, the Colonial Reports started to mention leprosy explicitly. In 1923, forty-three persons were examined in Suriname, of whom twenty were diagnosed with the contagion and thirteen labelled as suspect.[25] The next year, thirty-five out of sixty-five investigated persons were diagnosed with leprosy, with a further twenty suspects. In 1924, the Dutch Parliament asked for more energetic policies against leprosy.[26] The 1925 Colonial Report railed against the population's indifference and how sufferers lived undetected at home. In 1927, a public health inspector in the Dutch East Indies uttered his indignation in the *Dutch Journal of Medicine* that according to his information, the Paramaribo police had ignored the leprosy edict and did not believe in contagion. If agents arrested a sufferer in public after dark this led to a riot. A child's nurse in the family of a medical officer had been accused of leprosy, but the police had not reported her to the leprosy committee.[27] The Dutch medical doctor and public health officer Lampe claimed that there were approximately 1,000 leprosy sufferers running free in town (2.5 to 3 per cent of the town population), and warned that children ran the risk of contact and infection.[28] A new investigation in 1927 estimated that there was a total of 960 sufferers in the colony. Only 370 of them were admitted to asylums, and roughly the same number were under observation of the medical service in the town or the districts; however, 209 were 'missing'.[29] To the colonial government and Dutch doctors, action seemed necessary.

Dutch colonial medicine and the need for reform

In 1923, Lampe had arrived in Suriname where he would become instrumental in modernizing leprosy policies. He was born in 1895 and after his medical studies he joined the Military Medical Services of the Dutch colonial army as a (junior) health officer of the second class.[30] Although initially he was sent to Suriname for three years, he ended up staying for ten and became the public hygiene specialist for Surinamese healthcare.[31] His 1927 publication on the 'social hygiene' of the Surinamese population cemented his reputation.[32]

A close cooperation existed between Lampe's military medical service and the civil medical service in Suriname. Peters, the man sent on the leprosy fact-finding mission to the British Caribbean in 1890, was a military health officer as well as a member of the leprosy committee and a surgery and anatomy teacher at the Medical School.[33] After Salomons retired as medical inspector in 1908, this position remained vacant for financial reasons until 1921; in the meantime, the director of the military medical service conducted the medical inspections. Both functions were combined in 1927. In the modern colonial state, civil and military medical practitioners closely cooperated in disease prevention and healthcare for the civilian population. This is unsurprising in a colony where from the very beginning military and civil personnel had cooperated in importing and controlling a non-Western labour force. Military health officers were involved in healthcare for the poor. From 1889 onwards, the Medical School was located in the Military Hospital, which at the time was the only hospital in Paramaribo. The civil hospital (the Burgerziekenhuis) had burned down in 1821 and a new Catholic hospital was only built in 1917.[34] Contrary to some typologies of colonial medicine in other regions, where military considerations took precedence over civil healthcare, in Suriname there was no real division between military and civil healthcare.[35]

When the functions of director of the civil and military health services were combined in 1927, a special department for common and infectious diseases ('volks- en besmettelijke ziekten') was founded. Lampe was appointed head of the department and was given two assistants, a medical doctor who specialized in the field of leprosy and another who specialized in other infectious diseases, even though these appointments meant the budget was significantly exceeded by 18,000 guilders.[36] In May 1929, Lampe became medical inspector and could finally execute his plans for the new leprosy edict.[37] By then Lampe had become one of the great advocates of a new policy, together with F. P. Schuitemaker.[38] Schuitemaker had graduated from the Surinamese Medical School in 1913, and had continued his studies in the Netherlands during the First World War. In 1919, he returned to Suriname and set up private practice in Paramaribo.[39] He became a member of both the medical and (as Lampe had) the leprosy committee.[40] The ideas of doctors sent from the Netherlands and those native of the colony were congruent.

A Surinamese newspaper commented on the 'fanatical fight' of the civil health service against leprosy and how Lampe published statistical data to convince people that they would become extinct if the prevention of leprosy was not executed more rigorously.[41] At the same time,

Lampe came under severe criticism from the Surinamese newspapers for his policies and especially for his treatment of Javanese and British Indian immigrants, as well as for his seeming neglect of healthcare in the districts where these migrants lived.[42] His personal supervision of the repatriation to Java of 'dozens' of Javanese migrants who were no longer welcome in Suriname increased his unpopularity among migrants.[43] His decision not to accept a British Indian patient into the hospital for administrative reasons instead of giving precedence to humanitarian considerations became a scandal and was even discussed in the Dutch Parliament.[44]

By the time he arrived in Suriname, Lampe had been educated in the latest Western medical knowledge on 'tropical diseases'. This category was unknown before the end of the nineteenth century, since diseases that were prevalent in the tropics could also occur in non-tropical regions. However, the specialized study of tropical medicine had received a major impetus through the work of the Scottish physician Patrick Manson. Manson's medical textbook *Tropical Diseases: A Manual of the Diseases of Warm Climates* was published in 1898 and continued to function as a major textbook in revised editions. In the original manual, Manson included leprosy among those diseases that were especially prevalent in tropical climates.[45] Manson realized that leprosy could and did occur outside of these climates. Although he recognized that only two or three comparatively unimportant diseases strictly deserved the title 'tropical diseases', he treated leprosy as a disease specifically found in tropical and sub-tropical countries.[46] Basing his textbook on the new insights of bacteriology, Mason was convinced that the leprosy bacillus discovered by Hansen was the cause of the disease and that the Hansen bacillus was transmitted directly from one leprosy sufferer to the next (and not, for instance, by insects).[47] This idea made specific interventions possible that were not directed at changes in personal and social hygiene, and environmental conditions (now called 'lifestyle') as was characteristic for pre-modern medicine, but rather specifically aimed at a germ using Western medical technology. Historians have dubbed this kind of intervention 'Imperial Tropical Medicine': a medicine that intervenes 'from above' to eliminate the causes of diseases based on Western medical knowledge and imperial political institutions.[48]

There is another subtext in this new Imperial Tropical Medicine that is related to an older history of colonial medicine in slave societies. In this subtext, leprosy is identified as a disease of inferior people. For all his good intentions, Mason did approve of the idea that leprosy was a disease of 'semi-civilization'. For Mason,

> Savages are exempt, the highly civilised are exempt, but when the savage begins to wear clothes and live in houses he becomes subject to the disease. In other words, in the early stages of civilization, opportunities of infection are multiplied, and their influence is not counteracted by cleanliness of house or person.[49]

The notion of the robust health of the noble savage was a common one going back to the sixteenth century. Eighteenth- and nineteenth-century descriptions routinely denied the existence of leprosy among the Amerindians of Suriname with their healthy lifestyle and diet. In contrast, the 'semi-civilized' Afro-Surinamese were considered carriers of leprosy since the eighteenth century.

In the year that Lampe came to Suriname, a new Dutch textbook was published consolidating the latest state-of-the-art knowledge on tropical diseases. The chapter on leprosy advocated a strict prophylaxis: segregation with regular and compulsory treatment was thought to be the best method to fight leprosy. The authors deplored that the general policies against leprosy in tropical countries were too lax. As best practices, the compulsory segregation policies in the Philippines, the Hawaiian Islands, and Suriname were mentioned. Although the book was written by physicians from the Dutch East Indies, in their view it was the West Indies that had been most successful in the fight against leprosy. However, they thought that for pragmatic reasons it was not feasible to import these policies to the East Indies, since the population there was considered far too large for the Dutch to control.[50]

When Lampe set sail for Suriname, he would not have doubted the need for continuing compulsory segregation policies in Suriname. However, on the ground in Suriname he concluded that the Surinamese policies were not so effective. The situation did not seem to differ much from the days of Landrè. According to Lampe, Suriname was one of the most infected lands in the world and he evaluated existing leprosy policies as insufficient.[51] Other influential medical practitioners with experience in Suriname agreed, including former medical inspector J. W. Wolff, P. C. Flu, the Surinamese-born professor of tropical medicine and director of the institute of tropical medicine in Leiden, and

Paramaribo general practitioners, such as Schuitemaker and T. May. The last three were all born in Suriname and were graduates of the Medical School. For Wolff, two factors were essentially obstructing the execution of leprosy politics in Suriname: the Afro-Surinamese population was not afraid of contagion, and when leprosy occurred among families of the higher classes, the police left these families alone. Wolff also sounded a sexual note: in 1922, two Europeans were found to be afflicted with leprosy, and both of them had had sexual contact with Afro-Surinamese women.[52] Professor Flu returned to Suriname for a few months at the end of 1927 and sketched an alarming picture of the problem of leprosy in Paramaribo.[53] Schuitemaker urged stricter policies and more limitations on the outside world contacts of sufferers in the asylums.[54]

All these practitioners emphasized that segregation policies could not solely be a matter for the police. The sufferers should be treated as patients, not as criminals, their affliction seen as a disease and not as a vice, and treatments with medication, such as chaulmoogra that had been experimented with since 1905, should be made more available and attractive. For milder cases, such as patients with few open wounds, outpatient treatment should be offered. For all these doctors, it was clear that the segregation policies in Suriname had to be reformed but not abolished. Whether their background was serving as a military doctor from the Netherlands or a Surinamese-born civilian medical practitioner, these doctors held a similar viewpoint and wish for the modernization of leprosy policies.

A third alternative

A third alternative in the international discussion on the desirability of compulsory segregation arose in the 1920s. When the third international leprosy conference was held in Strasbourg in 1923, a new camp questioned the recommendations of the Berlin and Bergen conferences of 1897 and 1909. This camp did not doubt the contagiousness of leprosy and the role of the Hansen bacillus, or the advisability of sufferer segregation. However, this camp took a stance against the necessity of compulsory segregation. The British medical doctor, a retired major general in the British Indian Medical Service and prominent member of the Royal Society for Tropical Medicine and Hygiene founded in 1907, Sir Leonard Rogers, told the delegations in Strasbourg that 'compulsory isolation ... was only justifiable if the best possible treatment was provided for the interned leprosy patients'.[55] Together with the Christian British Mission for Lepers in India, Rogers created the British Empire Relief Association (BELRA). In nineteenth-century British India, the medical service had hardly cared or had been barely able to care for the Indian population. Here policies of compulsory segregation as in Suriname were far from realistic. Nonetheless, for Rogers the disease had to be managed both from the viewpoint of leprosy as an 'imperial danger' and Christian paternalism and compassion.[56] Furthermore, since 1920 he had worked with British pharmaceutical companies, including Burroughs, Wellcome & Co., in developing leprosy medication based on chaulmoogra (hydnocarpus oil). He had high hopes for this treatment but, as he wrote in retrospect twenty-five years later:

> The improved treatment was only of material value in comparatively early cases of leprosy. [But] the only method of control of the disease in general operation was compulsory segregation. Under that system early cases suitable for treatment were all hidden for fear of imprisonment for life.[57]

Rogers was not opposed to segregation in and of itself. In fact, he was rather enthralled by the results of American leprosy policies in the Philippines and the compulsory isolation of sufferers there in the Culion asylum, because he and the Christian Mission worried about sexual contacts between sufferers and applauded their segregation, which he believed diminished the danger of contagion to children.[58] However, the Calcutta school of leprosy of Rogers and his medical colleague and successor at the School of Tropical Medicine Ernest Muir advocated a new approach in those countries where policies of compulsory segregation were not feasible. Asylums should be made attractive like farm colonies where sufferers might work and be instructed in hygienic living. In this way, sufferers and their families would become willing to participate in a process of early detection, voluntary admittance and isolation, and (chaulmoogra) treatment.[59] The availability of medication, even when its efficacy remained doubtful, was considered important in persuading sufferers to report to medical authorities. After all, Rogers wrote in 1925, not every sufferer was 'ready' for segregation, with the suggestion that this readiness for segregation was a mark of civilization.[60] He complained in this context about 'poor and backward tropical races' that supposedly did not enthusiastically embrace colonial health policies.[61]

Under Rogers' influence the Strasbourg conference of 1923 accepted the necessity of segregation in endemic centres of leprosy, and the recommendation that isolation should be as humane as possible.[62] In Strasbourg, there were no Dutch representatives present. Leprosy specialists from Suriname had already failed to appear at the Berlin and Bergen conferences of 1897 and 1909, but in 1923 even representatives of the Dutch East Indies were not present in Strasbourg.[63] Nevertheless, the ideas of Rogers and BELRA were of particular appeal to doctors in Suriname where the colonial state wished to modernize and socially engineer the colonies, and where the old ideas of exclusion and exiling of leprosy sufferers to far-away asylums were giving way to their treatment, or rather their education as modern citizens.

Historians influenced by the work of Michel Foucault have placed these new ideas in a transition from a crude sovereign power to a modern disciplinary government, locating them in institutional spaces where the sufferers would be disciplined and discipline themselves.[64] However, before this kind of disciplinary regime could (or could not) be exercised in the leprosy asylums, the key question that had to be addressed was how to get the leprosy sufferers into the asylums. A possible model for Lampe was the reform undertaken in the Philippines where the concentration of sufferers in one asylum had fallen out of favour by the end of the 1920s.[65] In the Philippines, leprosy sufferers with few visible deformities, but who tested positive for the leprosy bacillus through the inspection of nasal discharge, were not automatically sent to an asylum. Instead, treatment in outpatient clinics was offered to milder cases. Early detection was put in the hands of an experienced leprosy specialist who also monitored the outpatient clinic treatments. Similar policy reforms were developed in British Guiana. Influenced by these ideas, Lampe proposed that the segregation policies in Suriname be revised along similar lines. He visited Rogers and received the 'great man's' blessing. However, for Lampe, compulsory admission to an asylum continued to remain a necessary policy instrument since he believed that for those who violated the leprosy laws, compulsory admittance would be beneficial.[66]

The edict of 1929

Lampe and Schuitemaker received political support for their ideas. In 1926 and 1927, members of the Dutch Parliament again complained about the leprosy situation in Suriname, now basing their complaint on Lampe's statements.[67] In April 1927, the Governor sent a new version of the edict based on Lampe's ideas to the Colonial Estates.[68] It would take until October 1929 before the new edict became law. The new edict combined the continuation of compulsory segregation polices with the medicalization of leprosy policies advocated by Rogers and others. In the first article of the edict, medical practitioners under the direction of the medical inspector were given medical supervision of all leprosy sufferers and those suspected of leprosy. Asylums had to be built in healthy places, and as completely isolated from their environment as possible. Letters and all other objects had to be disinfected before leaving the premises, and men and women had to be separated unless they were married (Article 3). The modern asylum had to be as unlike Batavia had been as possible. At the same time, detection and segregation were to be executed as rigorously as the edict of 1830 had provided for. Detection and investigation were put in the hands of a leprosy committee of five, appointed by the Governor and chaired by the medical inspector (Article 6). Leprosy sufferers not admitted to an asylum had to live isolated in their homes and were prohibited to appear in public. The committee could force sufferers to be admitted to an asylum, but not to undergo medical treatment against their will (Article 9). Their houses and possessions could be disinfected or destroyed if deemed necessary (Article 21). A full-time medical doctor was appointed to supervise disease detection, examination of schoolchildren, and identification of sufferers (Article 10). Doctors, district physicians, owners and house tenants were obligated to report sufferers (Articles 11–13). Civil servants and the police had to provide medical practitioners with full cooperation (Article 14). Newly born children of leprosy sufferers should be examined and after a negative test result separated from their parents directly after birth. Then, the children should continue to be regularly examined (Article 15). Jail sentences of up to one month and fines of up to one thousand guilders were intended for offenders (Articles 16–18).[69] The new edict received almost unanimous support in the Colonial Estates, although one member, the physician May, wondered why there were no similar edicts against much more contagious diseases such as tuberculosis or syphilis.[70]

The authoritarian hand of the colonial state: the crisis of the 1930s

The new leprosy politics of the 1930s and 1940s coincided with what has been described by historian Rosemarijn Hoefte as the political high tide of the culture of domination, the 'ultimate expression of authoritarian colonial rule' under Governor Johannes Kielstra (1933–1944).[71] Government by the colonial state was firmly authoritarian. Most of Kielstra's decisions were legalized by royal edicts and did not need approval by the Estates.[72] Kielstra's governorship followed the slow process of the colonial state's increasing involvement in society. Though Suriname was a Dutch colonial construction and ultimately created by the power of the state, in the age of slavery that state was hardly present on the plantations ruled by their owners. The reorganization of agricultural production after emancipation necessitated the state's active involvement with labour immigration. The countryside was organized into districts and there was increased state involvement in education and the expansion of medical services.

The economic crisis starting in 1929 hit Suriname hard, since it was dependent on export revenue. Export values declined and unemployment increased, while the immigration of Javanese labour migrants continued. Wages, salaries, and pensions were cut, and taxes rose. According to Hoefte, 'Contemporaries stated that hunger, malnutrition, health problems, pollution, and dilapidation of houses and buildings were soon noticeable in Paramaribo ... In the districts, the situation was only slightly better.'[73] The crisis led to major social unrest, protests, and riots in the first half of the 1930s. As Hoefte writes, Suriname was a culture of domination and contestation:

> Contestation ... took many guises and ranged from cultural resistance by smallholder communities, which not only established an economic alternative to the plantations but also a more autonomous way of life by, for example, resisting religious efforts to impose European family forms ... to open political resistance in the forms of mass protests against administrations.[74]

The colonial state reacted to social unrest with more repressive policies. The police received new weapons and motorcycles, and intensified its intelligence gathering on potential sedition within the population. In 1933, freedom of the press and freedom of assembly were limited. Civilian patrols were organized to assist the military and police in quelling riots. It might be questioned how effective these policies were, since at the same time, budgets were cut and the number of policemen actually decreased from 231 in 1931 (of whom 145 were stationed in Paramaribo) to 197 in 1939 (129 in Paramaribo).[75] Nevertheless, the introduction of increased surveillance and detection of leprosy sufferers in 1929 has to be understood within this rather grim political, social, and economic climate.

Leprosy detection in action: the firm hand of the state

In reports to the Dutch Parliament, the colonial government emphasized its strict hand in enforcing the new leprosy politics even when recognizing the need for budget cuts because of the economic crisis as directed from the Netherlands. However, the Colonial Estates were opposed to any cuts in leprosy care. In May 1931, they voted against cuts in asylum budgets.[76] Against the wishes of the Dutch Minister of Colonies, the Estates granted a special gratuity to the teachers in the outpatient school for children with or suspected of leprosy that was opened that year. They also voted with a slight majority for the leprosy specialist to be provided with a special salary in the edict of 1929, against the wishes of the Dutch government once again. However, this last measure did not receive the necessary approval of the Dutch Parliament.[77] Private funding remained essential to keep the system of leprosy care going, not only in the Christian asylums, but also in the children's outpatient school that was opened in January 1931 and where food was provided by private funding.

Despite these financial problems, the colonial government presented the reorganized fight against leprosy as a success story. The Colonial Report of 1929 claimed that by then 90 per cent of all non-segregated sufferers outside of the asylums were under regular treatment.[78] The sufferers were taken from the streets of Paramaribo and delivered to hospitals and asylums.[79] In the districts, the fight against leprosy was catching up: in 1930, sixty-three people were diagnosed with leprosy, forty-six in 1931, and thirty-four in both 1932 and 1933.[80] Already in 1927, an outpatient clinic in Paramaribo with its own laboratory for diagnosis was opened.[81] Children with leprosy who lived with their parents were separated from other schoolchildren and from 1931 onwards received their education in the outpatient school.[82] The school had a capacity of 300 children.[83] In 1931, 165 children were admitted to the school but remained under medical observation. In that year, fourteen children's leprosy became too advanced to stay in the school, thus eleven were sent to asylums, and three were confined to their family homes. Of the other 151 children, five were discharged and sent to a normal school.[84] In the first ten years of its existence, a total of 375 children attended the outpatient school.[85] The existence of a school for children with leprosy in Paramaribo might suggest that the city population's fear of contagion was not great, since these children walked the streets in public. In addition, the school had no problem attracting a sufficient number of teachers despite the meagre salary of fifty guilders a month.[86] However, this might have had as much to do with the unemployment and the economic crisis at the time as with idealism or a lack of fear of contagion.

In 1931, for the first time, the Colonial Report was optimistic about the fight against leprosy. It said that the 'virus-reservoir', meaning the carriers of the leprosy bacillus, 'was more and more known and adapted.' This remark depersonalized sufferers into bacillus reservoirs and has quite a cold and mechanic ring. 'In the future prevention and treatment will be expected to have a very positive effect on the prevalence of the disease', continued the Report.[87] By 1935, the Public Health Service declared that the level of leprosy remained stationary.[88] This implied that in a period of only six to seven years, Dutch colonial medicine had dammed the growing danger of leprosy by using the modern policies of detection and treatment. If this was true, then it was a success story indeed. However, was this claim in accordance with the facts?

Detection in action: counting the numbers

What do the figures on the detection and segregation of leprosy sufferers show? Figure 3 and Table 3 demonstrate that from 1929 until 1940 the number of sufferers in the asylums grew significantly from a little more than 400 up to approximately 600. In the Second World War, a

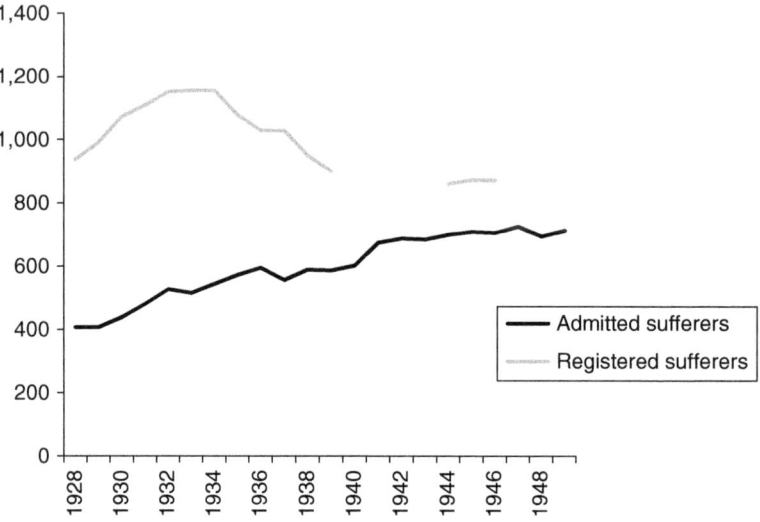

Figure 4 Admitted and registered leprosy sufferers, 1928–1949

second significant growth set in, bringing the total number up to over the 700 mark by 1945. From 1928 onwards, not only were the admitted leprosy sufferers registered, but also those who were segregated at home and the patients of the outpatient clinic. Figure 4 and Table 4 show the available figures compared to the asylum population.

What is remarkable is that until 1935 the number of registered sufferers was always more than twice the number of admitted sufferers, amounting to more than 0.7 per cent of the total population. After 1935, the difference between admitted and registered sufferers became smaller. The elderly registered sufferers died and during the Second World War more registered sufferers seem to have been transported to the asylums. While the percentage of asylum patients in the total registered population (that is, excluding the Maroons and the Amerindians) rose slightly during the war (from 0.38 per cent in 1939 to 0.41 per cent in 1946), the percentage of registered sufferers, including those outside the asylums, sank from 0.58 per cent in 1939 to 0.49 per cent in 1946, which was far below the level of 1929. When the percentages are recalculated according to the results of the more reliable 1950 census, the

Table 4 Admitted and registered leprosy sufferers 1928–1949

Year	Admitted sufferers	Registered sufferers
1928	408	938
1929	407	991
1930	439	1072
1931	482	1109
1932	527	1152
1933	516	1155
1934	544	1156
1935	573	1078
1936	595	1029
1937	557	1029
1938	588	950
1939	587	901
1940	601	
1941	675	
1942	688	877
1943	685	
1944	700	861
1945	709	873
1946	706	872
1947	726	
1948	695	
1949	714	

Source: Colonial Reports; GS 1262; Report Leprabestrijdingsdienst 1946, GS 1264.

picture remains the same. According to this recalculation, the asylum population grew between 1928 and 1946 by 0.1 per cent of the population, while the registered number of sufferers dropped in that period from 0.82 per cent to 0.45 per cent (see Tables 5 and 6; the percentages are higher using 1950 data since the recalculated general population figures are lower).

By the end of the 1940s, most registered leprosy sufferers were admitted to asylums. Their number had stopped rising even before

Table 5 Percentages of admitted and registered leprosy sufferers among the population, calculated according to the 1921 census

Year	Population (excl. Maroons and Amerindians)	Admitted to leprosy asylums	% Population (excl. Maroons and Amerindians)	Registered as leprosy sufferer	% Population (excl. Maroons and Amerindians)
1929	131,687	407	0.31	938	0.71
1935	146,843	573	0.39	1,078	0.73
1939	156,332	601	0.38	901	0.58
1946	177,880	726	0.41	872	0.49

Source: Lamur, H. E., 'The Demographic Evolution of Surinam 1920–1970: A Socio-Demographic Analysis' (PhD thesis, University of Amsterdam, 1973), p. 136; Colonial Reports.

Table 6 Percentages of admitted and registered leprosy sufferers among the population, recalculated according to the 1950 census

Year	Population (excl. Maroons and Amerindians)	Admitted to leprosy asylums	% Population (excl. Maroons and Amerindians)	Registered as leprosy sufferer	% Population (excl. Maroons and Amerindians)
1929	114,516	407	0.36	938	0.82
1935	129,672	573	0.44	1.078	0.83
1939	139,161	601	0.43	901	0.65
1946	160,709	726	0.45	872	0.54

Source: Lamur, 'Demographic Evolution', p. 136; Colonial Reports.

the introduction of effective medication. On the basis of these figures, one might credit Dutch colonial medicine with halting the spread of leprosy, if one accepts two premises. One is that before 1929 the incidence of the disease was still rising, a claim for which there is no reliable evidence. The second premise is that the detection system of leprosy sufferers was effective. However, both are uncertain. In the 1950s, the incidence of leprosy increased for a number of reasons, including the migration of Maroons (who in the period of direct Dutch colonial rule had not been under surveillance) within Suriname. In 1963, the total number of leprosy sufferers was 1,183, and there were 2,687 persons in whom the course of the disease was arrested.[89] In the early 1970s, just before Suriname became independent, more than 200 new cases of leprosy were discovered each year. Estimates of the prevalence of leprosy were as high as 2 per cent in 1970.[90] Though methods of diagnosis had changed and the migration among Maroons influenced the composition of the population, the possibility remains that many carriers of the Hansen bacillus had not been found by 1950. This more negative evaluation of the modern leprosy politics is supported by the many references in the colonial sources to the population's lack of cooperation.

Infrapolitics and leprosy: non-cooperation and friction

Without a doubt there were many medical, social, and financial benefits of the new politics for leprosy sufferers and their families. However, these were politics that were decreed and executed in a top-down fashion. When we consider the Suriname of the 1930s as characterized by crisis, unrest and contestation, it is unsurprising that such top-down politics of compulsory segregation was not always welcomed by the whole population.

In 1930, there already was a gender conflict. Women protested against their subjection to repeated examinations by medical doctors who were unknown to them, thus contesting the policies of medical authorities. The edict of 1929 decreed that the leprosy specialist who assisted the leprosy committee had to make regular examinations of all schools at least once in every four years.[91] There was a logistical problem, however. The leprosy specialist was not capable of doing all the examinations himself, especially when he had to travel to the districts. Since medical personnel changed quickly in these districts, this meant that teachers and pupils in the district schools had to submit themselves to intimate examinations by people who were unknown to them and who were often relatively young, inexperienced and sometimes had not yet graduated from the Medical School. This rankled female teachers who had to be examined and they protested. One teacher refused to be examined by such a doctor. Many others sympathized with her and the case caused quite a stir. To ensure cooperation with medical examinations the colonial government proposed to amend the leprosy edict, making it possible for people to ask for an examination by another doctor. In December 1930, this amendment was approved in the Colonial Estates by a small majority of seven to five votes after the medical inspector had warned that without approval leprosy politics were endangered.[92]

The problem of a lack of qualified physicians to examine the leprosy sufferers continued to exist. To make a conclusive examination, and to give the final diagnosis of leprosy with its far-reaching consequences, the leprosy edict decreed that a meeting of the entire leprosy committee was required, with all five members present. To ensure the feasibility of this, there were two deputy members appointed who could fill in for other members. In 1941, it was decided (possibly on the initiative of Schuitemaker who had by then become medical inspector) that the number of members of the leprosy committee was too small, which made it difficult to summon a complete session. Therefore, in July of that year, the number of deputy members was doubled to four.[93]

The population was not completely docile when faced with the new top-down medical interventions in their lives; teachers were not alone here. The Colonial Report of 1934 complained that the new leprosy edict was not supported by the good faith and the views of the population. Two-thirds of those who were ordered to appear before the leprosy committee for examination disappeared. Many continued their professional lives, mingled with the general population, and travelled by public transport. When they were finally caught and sent to the government asylum Groot-Chatillon, they escaped and lived hidden in the districts. According to the Report, the population was not sympathetic to the leprosy policies and the police force was too small to enforce the edict.[94] The difficulties in executing the leprosy edict might have been a factor in the resignation of E. G. Keil, the German leprologist trained at the Institute of Tropical Medicine in Hamburg, in October 1935.[95]

Sometimes the Surinamese population's discontent was heard in the Dutch Parliament or the Colonial Estates. The conclusions of the Colonial Report of 1934 led to alarm in the Dutch Parliament. The left-wing opposition, the Social Democrats and the Communists, used the Surinamese population's criticism in their attacks on the Dutch government. Agnes De Vries-Bruins, physician and Social Democratic spokeswoman for medical affairs in Parliament, was in the forefront of a clamour for stricter politics and execution of the 1929 edict. In discussions on the colonial budget of 1935, she supported the ideas behind the leprosy edict, but at the same time she criticized the execution.[96] In February 1936, De Vries-Bruins expressed her fears that the new leprosy politics in Suriname would fail. She accused the colonial authorities of lacking tact when managing the population. She rejected the role of the police, untrained in medical affairs, in rounding up leprosy sufferers and suspects. A district physician in Nickerie had sent forty-six people to Paramaribo to be examined by the leprosy committee, with the possible result that they would be sent from the examination to an asylum without ever seeing their family again or saying farewell. For De Vries-Bruins, this was inhumane, as was the story of a boy who could not go to school because he was suspected of leprosy. He was only examined again after four years and then freed of suspicion. Although De Vries-Bruins fulminated against the excesses of the system, it was unclear what her alternative proposition was. She rejected the system of compulsory segregation at the same time as she applauded the edict of 1929. She wished for a more efficient execution of the edict by medical personnel at the same time that she deplored people being segregated in the asylums as regulated by the edict. The Social Democratic spokeswoman gave voice to discontent in the Dutch Parliament, but she was not an opponent of the new system of leprosy politics per se.[97] In 1939, just before the outbreak of the Second World War, which severed the connections between Suriname and the German-occupied Netherlands, she applauded the leprosy politics in Suriname for having reduced the total number of leprosy sufferers, while complaining that 'difficult' and unruly elements had to be admitted to asylums. Once again, De Vries-Bruins did not come up with a constructive alternative.[98] Nor did the Communist leader David Wijnkoop have an alternative, although he stated in Parliament that all sufferers had to be treated on an individual basis as much as possible and recognized that more strict measures against sufferers were inevitable, thus showing his sympathies for the sufferers as well as for the people who had to manage them.[99]

The Dutch government denied the accusations of the left-wing opposition. In response to De Vries-Bruin's accusation, Hendrik

Colijn, the Prime Minister and Minister of Colonies, explained in 1936 that suspects had to be examined by the complete committee and laboratory facilities for an adequate diagnosis were only possible in Paramaribo, which was the reason that suspects were sent there. The people concerned in the Nickerie affair had been quite satisfied with their treatment according to the Minister's information. And the boy mentioned by De Vries-Bruins was, Colijn claimed, still under regular control by a physician, and had not been cleared from suspicion at all.[100]

By this time, discontent about a number of other elements of the execution of the new leprosy politics had also been voiced in the Colonial Estates. In December 1937, the Governor proposed to the Estates that there should be a change in the leprosy edict that gave the leprosy physician and the district physicians the authority to deal with matters of transport of suspected leprosy sufferers, decontamination of their dwellings, and the decontamination and, if possible, destruction of their possessions.[101] The Estates' committee of report in March 1938 expressed the sufferers and their families' discontent about these measures. Who should pay for and organize the sufferers' transport to Paramaribo when they had to be examined by the leprosy committee? And what should happen to their possessions? In principle, their possessions were possibly contaminated and should be destroyed. However, in practice they were sold or ended up in government institutions. One particular focus of discontent was the fate of the piano belonging to a person suspected of leprosy. The piano ended up in the national (former military) hospital in Paramaribo, causing suspicions that officials turned the situation of the leprosy sufferers to their own advantage and that declaring possessions to be contaminated was only a subterfuge in order to achieve this end. A second source of discontent was the lack of financial support for the outpatient school. Private funding organizations had to step in to make sure there was adequate food and nourishment for the children who had been compulsorily sent there. A third source of discontent was the fate of those dismissed from the asylums, which remained uncertain. The execution of leprosy politics was deeply entwined in the lives of the social communities and could not be left solely to the physicians appointed by the state.[102] Furthermore, the committee's report asked for private physicians to participate in the activities of the leprosy committee.[103]

Faced with this criticism, the colonial authorities immediately returned the piano to its rightful owner.[104] However, the problem of the discharged patients and their rehabilitation in society was not so easy to solve. The medical inspector, P. Cool, explained to the Estates that these people were often returned to their families and friends only to find that there was no place for them any more. Although they were now free of the bacillus, there was a saying among the people that 'once a leper always a leper'.[105] Thus, former patients wished to return to the asylum. In Cool's view, the former patients should remain suspects. He suggested a project to segregate the former leprosy sufferers in special dwellings, but admitted that they could not be forced to go there. He also defended himself about the piano, for which he was responsible, and promised that he would advise suspects not to sell their possessions. Finally, the Estates supported the proposed change in the edict and government policies.[106]

How can one understand this discontent and non-compliance with the leprosy edict among the population? Many sufferers and their families were quite ready for medical assistance and treatment. The problem of the rehabilitation of former leprosy patients from the asylums who were not accepted by their families upon return shows that it was not simply a matter of *them* (the colonial authorities and colonial medicine) against *us* (the sufferers and their families). But this should not suggest that the population was essentially docile. There was clearly a lack of cooperation, as well as resentment and friction.

Resistance always stands in a relationship with domination. When applying James C. Scott's analytical scheme of domination and resistance to the leprosy situation in the interwar period in Suriname, practices of material domination, status domination, and ideological domination by the colonial rulers are noticeable. The sufferers and their children, as well as their dwellings and possessions, were appropriated in a process of material domination. The sufferers were unprivileged and lost their status in society, a move ideologically justified on medical and public health grounds. In opposition to this domination, there were forms of low-profile and undisclosed resistance in everyday life, or what Scott calls 'infrapolitics'. In Suriname, this resistance included a refusal to cooperate with examinations, and hidden transcripts of anger about treatment and disrespect (because of stigmatizing, or forced transports to Paramaribo). As the following chapters show, people held fast to their own folk beliefs about the supernatural influences in the aetiology of leprosy, despite the ideas of Western-educated doctors, and patients attempted to escape temporarily or permanently from the asylums.

The Second World War and the end of Dutch colonial rule

In Suriname, the Second World War was a period of economic improvement, especially after the United States entered the war. In 1942, 2,000 American soldiers were garrisoned in Suriname to protect the bauxite mines against possible Nazi sabotage since they were vital for the Allied war effort. Activities in the bauxite industry intensified, leading to a fivefold increase in employment in the sector. Employment also rose because of the construction of defensive works, roads and the expansion of the Zanderij airport. The American soldiers caused a rise in consumer spending and an expansion of the service sector. Thousands of plantation workers, especially British Indians and Javanese, left for Paramaribo to find work for relatively high wages, while the income of small farmers also rose. Only those with a fixed income, such as civil servants, saw their spending power decrease because of inflation.[107]

This unexpected economic boom had its effect on the execution of leprosy policies. Looking at the figures for the war period in Figure 3 and Table 3, at the beginning of the war the focus on leprosy seems to have almost disappeared. The influx of American soldiers in Suriname and the resulting social, economic, and demographic changes detracted from the focus on the fight against not only leprosy but also against other contagious and endemic diseases. However, this changed again by the end of the war. In 1944, the department of public hygiene, founded by Lampe at the end of the 1920s, was reactivated, which suggests that at some point during the war its activities had almost ceased. A public health service report stated that an active fight against diseases was once again needed.[108] Although leprosy was not specifically mentioned in this report, surveys of leprosy sufferers were held again, and in 1944 the leprosy committee performed more examinations than ever before: 650 in total, in forty-four sessions.[109] This number was almost the same in 1946: 644 examinations, in forty-four sessions.[110] In 1946/47, a report of the Dutch parliamentary committee for the West Indian colonies, installed after the liberation of the Netherlands from German occupation, asked for extensive funding of leprosy research in Suriname.[111] At that time, sulfone therapy was introduced in Suriname, so that for the first time doctors in and outside the asylums had an effective medication against leprosy. However, the introduction of the medication was not without problems, especially because of its high cost. Patients with leprosy in the government asylum Groot-Chatillon protested when they heard about the new medicine that they were not administered.[112] Private funding was organized to meet the financial costs. There was support from the United States and in 1950 the Dutch Red Cross, with private support, financed the production of 500,000 DDS (diaminodiphenyl sulphate) tablets for the governor of Suriname, with the promise of more.[113]

By this time, Dutch control of leprosy policies in Suriname was coming to an end. Forced by the Atlantic Charter agreed upon by Roosevelt and Churchill in 1941 that had promised self-determination for all nations if the Allies won the war, and the alliance of Indonesian nationalists with the Japanese, the Dutch Queen Wilhelmina promised self-government in internal affairs to all Dutch colonies in a radio speech in December 1942.[114] After the war, this promise had to be kept, especially because of the United States' dominance. In 1949, the first general election with universal suffrage in Suriname was held. From 1 January 1950, internal affairs became the responsibility of the chosen Surinamese government that would not have to account to the Dutch government in the Netherlands for its policies, rather only to the chosen Estates of Suriname (the former Colonial Estates). This transfer of Dutch control over Surinamese internal affairs foreshadowed the treaty of 1954 in which Suriname received a semi-autonomous status within the Kingdom of the Netherlands and its overseas territories. Twenty years later, Suriname became an independent country in 1975.

The problem of leprosy was not yet completely solved, but the availability of effective medication led to the end of two centuries of compulsory segregation in Suriname. The outpatient clinic school was closed in 1958, Majella in 1964, Bethesda in 1968, and finally

Groot-Chatillon in 1973 when compulsory segregation was abolished. Since 1953, the private Esther Foundation has been active in the housing, care, and reintegration of patients into society. However, leprosy has not disappeared. By the beginning of the twenty-first century, on average, fifty new patients were being diagnosed with the disease each year. Nevertheless, leprosy's importance in public health politics has diminished.[115] Effective treatment and multi-drug therapy are available and DDS is on the World Health Organization (WHO) list of essential drugs available for basic health systems. Once perceived as a threat to colonial empires, leprosy has become a so-called NTD – a neglected tropical disease.

Conclusion

After the reorganization of the leprosy asylum system, the Surinamese colonial government directed its attention to the reform of the legal framework of compulsory segregation that still dated from the period of slavery. In 1904, it presented a draft for a new leprosy edict that combined increased surveillance and detection and increased isolation of the asylums with an improvement in medical provisions. This edict was based on a fear of increasing leprosy, especially among the non-white working classes, the Afro-Surinamese and to a lesser extent British Indians. However, when a survey in 1908 indicated that leprosy might not be on the increase at all, the urgency disappeared from the project. This urgency returned in the course of the 1920s, driven by an alliance between Dutch military doctors sent from the Netherlands and Suriname-born civilian doctors. This group embraced the ideas of a 'third way' for leprosy care, a combination of compulsory segregation with humane treatment including medication and outdoor care. These ideas fitted well into authoritarian modernist colonial policies, and led to the new leprosy edict of 1929. From this time forward, the last two decades of Dutch colonial rule witnessed the most complete execution of modern leprosy politics, albeit with a Janus head. On the one hand, the edict of 1929 emphasized and stimulated medicalization of leprosy, the possibility of medical treatment in asylums, at home and in an outpatient clinic, and a humane organization of life in the asylums according to the standards of the day. On the other hand, policies of surveillance, detection, and compulsory segregation were intensified. This

Developing modern leprosy politics, 1900–1950

side showed that the leprosy politics in the last decades of direct Dutch colonial rule of Suriname were still characteristic of the general 'culture of domination' in the colony. The modern leprosy politics of the 1930s and 1940s were executed in the heyday of authoritarian colonial rule. Furthermore, the policies were still based on the stigmatization of Afro-Surinamese and British Indian sufferers and on the perceived need to force them to comply with health measures. The Surinamese contested these authoritarian policies. An infrapolitics of non-cooperation and non-compliance remained in force against the domination of colonial government and colonial medicine.

Notes

1. J. C. Scott, *Seeing Like s State: How Certain Schemes to Improve the Human Condition Have Failed* (New Haven, CT: Yale University Press, 1998), pp. 87–102.
2. Scott, *Seeing Like a State*, pp. 87–102.
3. D. Arnold, *Colonizing the Body: State Medicine and Epidemic Disease in Nineteenth-Century India* (Delhi: Oxford University Press, 1993).
4. Colonial Reports (hereafter CR).
5. T. Broes van Dort, 'Een en ander over de lepra in Nederland en zijne koloniën', *Nederlandsch Tijdschrift voor Geneeskunde* 41 (1897), pp. 292–6, 384–91, 407–21, 650–1, on pp. 407–21; T. Lens, 'Lepra in Suriname', *Elsevier's Geïllustreerd Maandschrift* 10 (1895), pp. 521–52, on p. 527.
6. As quoted in *Handelingen Staten-Generaal*: www.statengeneraaldigitaal.nl/ (hereafter SG), 19 December 1896.
7. National Archive, The Hague, Handelingen en Bijlagen van de (Koloniale) Staten van Suriname (hereafter CE), 1904/1905, Bijlagen 10.3.
8. P. C. Flu, *De filaria-ziekte in Suriname* (The Hague: Algemeene Landsdrukkerij, 1911).
9. CE 1904/1905, Bijlagen 10.2, 10.3.
10. CE 1905/1906, Bijlagen 24.1.
11. CE 1906/1907, Bijlagen 1.3; 1.4.
12. CE 1906/1907, pp. 12–19: session Colonial States 6 August 1906.
13. CE 1908/1909, Bijlagen 23.1.
14. CE 1909/1910, Bijlagen 168/20.
15. CE 1908/09, Bijlagen 23, 1–3.
16. CE 1908/09. Bijlagen 14, 1–3.

17 CE session Colonial States 5 September 1917.
18 CE 1916/1917, Bijlagen 1/1.
19 CE 1919/1920, Bijlagen 3.1, 3.2.
20 CE 1911/1912, Bijlagen 157/15.
21 CE 1917/1918, Bijlagen 1/4.
22 CR 1918, p. 3.
23 CE 1915/1916, Bijlagen 1/5.
24 C. Bonne, 'De maatschappelijke beteekenis der Surinaamsche ziekten', *West-Indische Gids* 1 (1919), pp. 291–300.
25 CR 1924, Appendix H.
26 CE 1923/1924, Bijlagen 1/2.
27 P. Cool, 'Regeeringsmaatregelen ter bestrijding der lepra in Suriname en Aruba', *Nederlandsch Tijdschrift voor Geneeskunde* 71 (1927), pp. 2453–6.
28 CR 1925, pp. 19–20; CE 1925/1926, SG: Handelingen Tweede Kamer 20 April 1926; P. H. J. Lampe, 'Sociaal-hygiënische beschouwingen' (Kon. Vereeniging Koloniaal Instituut, Mededeeling no. XXIII, 1927), pp. 237–8.
29 CR 1928.
30 B. C. de Jonge, *Herinneringen* (Groningen: Wolters–Noordhoff, 1968), p. 485.
31 *Suriname* 15 June 1923; *De West* 31 May 1933; www.delpher.nl/nl/kranten/ (hereafter SN); *Vraagbaak. Almanak voor Suriname 1928* (Paramaribo: Van Ommeren, 1927), p. 123, www.dbnl.org/ (hereafter SA).
32 P. H. J. Lampe, 'Sociaal-hygiënische beschouwingen' (Kon. Vereeniging Koloniaal Instituut, Mededeeling no. XXIII, 1927).
33 SA: *Surinaamsche Almanak voor het jaar 1891* (Paramaribo: Erve J. Morpurgo, 1890), pp. 150–2.
34 J. Karbaat, 'De historie van de militair-geneeskundige dienst in Suriname', *Nederlands Militair Geneeskundig Tijdschrift* 17 (1964), pp. 275–7; G.-J. Hallewas, 'De gezondheidszorg in Suriname' (PhD thesis, Groningen University, 1981), pp. 87, 96. In 1934 the Military Hospital became a national civil hospital ('s Lands Hospitaal). Between 1942 and 1944 a separate new military hospital was constructed.
35 Harrison, M., 'The tender frame of man: Disease, climate, and racial differences in India and the West Indies', *Bulletin of the History of Medicine* 70 (1996), pp. 68–93; A. H. M. Kerkhoff, 'The organization of the military and civil military service in the nineteenth century', in G. M. van Heteren. A. de Knecht-van Eekelen and M. J. D. Poulissen (eds.), *Dutch Medicine in the Malay Archipelago 1816–1942* (Amsterdam: Rodopi, 1989), pp. 9–24.

36 Hallewas, 'Gezondheidszorg', pp. 86–7; SG: Tweede Kamer 1928–1929 kamerstuknummer 415 ondernummer 3; 1930–1931 Kamerstuk 84 ondernummer 1, 2.
37 SG: Tweede Kamer 1930–1931 Kamerstuk 428 ondernummer 3.
38 SA: *De Vraagbaak. Almanak voor Suriname 1928*, p. 125.
39 SN: *Suriname* 18 April 1913; *De West* 21 January 1919, 2 April 1919.
40 Functions Schuitemaker and Lampe: SA: *De vraagbaak, Almanak voor Suriname 1925* (Paramaribo: Van Ommeren, 1924), pp. 364–5; *De vraagbaak, Almanak voor Suriname 1928* (Paramaribo: Van Ommeren, 1927), pp. 362–3.
41 SN: *Suriname* 31 December 1929.
42 SN: *Suriname* 29 December 1929; *Suriname* 19 September 1931;*De banier van waarheid en recht* 7 June 1932.
43 CR 1931, p. 2.
44 SN: *Suriname* 31 March 1931.
45 P. Manson, *Tropical Diseases: A Manual of the Diseases of Warm Climates*, rev. ed. (London: Cassell, 1903 [1898]), p. xix.
46 Manson, *Tropical Diseases*, pp. xxiv, 480–1.
47 'Modern science has clearly shown that nearly all diseases, directly or indirectly, are caused by germs.' (Manson, *Tropical Diseases*, p. xx.)
48 M. Worboys, 'The emergence of tropical medicine: A study in the establishment of a scientific specialty', in G. Lemaine, R. Macleod, M. Mulkay, and P. Weingart (eds.), *Perspectives on the Emergence of Scientific Disciplines* (The Hague: Mouton, 1976), pp. 75–98; J. Farley, *Bilharzia: A History of Imperial Tropical Medicine* (Cambridge: Cambridge University Press, 1991); D. M. Haynes, *Imperial Medicine: Patrick Manson and the Conquest of Tropical Disease* (Philadelphia, PA: University of Pennsylvania Press, 2001).
49 Manson, *Tropical Diseases*, p. 508.
50 C. D. de Langen and A. Lichtenstein, *Leerboek der tropische geneeskunde*, 2nd rev. ed. (Weltevreden: G. Kolff, 1928), p. 234.
51 P. H. J. Lampe, 'Enkele opmerkingen over den sociaal-hygiënischen toestand en de geneeskundige verzorging van Suriname', *West-Indische Gids* 8 (1927), pp. 249–76, on pp. 257–8, 272.
52 J. W. Wolff, 'Het lepra-probleem in Suriname', *Geneeskundig Tijdschrift voor Nederlandsch-Indië* 65 (1925), pp. 572–87.
53 G. A. Lindeboom, *Dutch Medical Biography: A Biographical Dictionary of Dutch physicians and surgeons 1475–1975* (Amsterdam: Rodopi, 1984), p. 597; P. C. Flu, *Verslag van een studiereis naar Suriname*

(*Nederlandsch Guyana.*) *Sept. 1927 – Dec. 1927* (Utrecht: Kemink, 1928), pp. 65–88.
54 SN: *De West*, 2 July 1926, BP T 153.
55 L. Rogers, *Happy Toil: Fifty-five Years of Tropical Medicine* (London: Frederick Muller, 1950), p. 196.
56 M. Worboys, 'The colonial world as mission and mandate: Leprosy and empire, 1900–1940', *Osiris* 15 (2000), pp. 207–18.
57 Cit. in T. Gould, *Don't Fence Me In: Leprosy in Modern Times* (London: Bloomsbury, 2005), p. 223.
58 Gould, *Don't Fence Me In*, p. 215.
59 Worboys, 'Colonial world'; M. Vollset, 'Globalizing Leprosy: A Transnational History of Production and Circulation of Medical Knowledge 1850–1930' (PhD thesis, University of Oslo, 2013), p. 154–7.
60 Vollset, 'Globalizing Leprosy', p. 156.
61 Cit. in Gould, *Don't Fence Me In*, p. 222.
62 In Vollset, 'Globalizing Leprosy', p. 310.
63 J. D. Käyser, 'Beschouwingen naar aanleiding van het Verslag van de IIIde Internationale Lepraconferentie, gehouden te Straatsburg van 28–29 Juli 1923', *Geneeskundig Tijdschrift voor Nederlandsch-Indië* 65 (1925), pp. 716–50, on p. 717.
64 A. Bashford, *Imperial Hygiene: A Critical History of Colonialism, Nationalism and Public Health* (Basingstoke: Palgrave Macmillan, 2004), pp. 89–93; W. Anderson, *Colonial Pathologies: American Tropical Medicine, Race and Hygiene in the Philippines* (Durham, NC: Duke University Press, 2006), pp. 156–79; Vollset, 'Globalizing Leprosy', p. 157.
65 Anderson, *Colonial Pathologies*, p. 179.
66 Lampe, 'Sociaal-hygiënische beschouwingen', pp. 526–31.
67 CE 1927/1928, Handelingen Tweede Kamer, 23 February 1927, Bijlagen 1/1; CR, p. 23.
68 CE 1926/1927, Bijlagen 41.1–3.
69 *Gouvernementsblad* 1929, No. 79, 25 October 1929.
70 CE 1928/1929, pp. 83–99.
71 R. Hoefte, *Suriname in the Long Twentieth Century: Domination, Contestation, Globalization* (New York: Palgrave MacMillan, 2014), p. 91. For Kielstra as Governor see H. Ramsoedh, 'Suriname 1933–1944. Koloniale politiek en beleid onder gouverneur Kielstra' (PhD thesis, Utrecht University, 1990).
72 Ramsoedh, 'Suriname', pp. 133–45.
73 Hoefte, *Suriname*, p. 71.
74 Hoefte, *Suriname*, p. 2.

Developing modern leprosy politics, 1900–1950   197

75 E. Klinkers, *De geschiedenis van de politie in Suriname, 1863–1975. Van koloniale tot nationale ordehandhaving* (Amsterdam: Boom, 2010), pp. 105–28 (numbers of police officers on p. 128); Hoefte, *Suriname*, pp. 69–90.
76 Archive Bisdom of Paramaribo, T 177.
77 SG: Handelingen Tweede Kamer 25 February, 1 April 1932.
78 CR 1929, p. 3.
79 CR 1929, p. 16.
80 National Archive, Paramaribo, Gouvernementssecretarie Suriname 1829–1954 (hereafter GS), nr. 1248.
81 Hallewas, 'Gezondheidszorg', p. 43.
82 CR 1931, p. 18.
83 SN: *Suriname* 20 May 1930.
84 National Archive, Paramaribo, Gouvernementssecretarie Suriname 1829–1954 (hereafter GS), nr. 1258.
85 Hallewas, 'Gezondheidszorg', p. 124.
86 SN: *Suriname* 20 May 1930.
87 CR 1931, p. 18.
88 CR 1935, p. 59; CE 1935/1936, Tweede Kamer Bijlagen 1/5.
89 Kuyp, E. van der, 'De geschiedenis van lepra in Suriname tot 1971', *Surinaams Medisch Bulletin* 14 (1999), 2, pp. 36–55, on p. 52.
90 Hallewas, 'Gezondheidszorg', p. 41.
91 CE 1930/1931, Bijlagen 14.4, 27 November 1930.
92 CE 1930/1931, pp. 113–14 (session 18 December 1930); Bijlagen 14.1, 14.2, 14.4, 14.
93 CE 1941/1942, sessions 15 May, 12 June, 10 July 1941.
94 CR 1934, p. 50.
95 Hallewas, 'Gezondheidszorg', pp. 123–4; BP T 186; BP T 187.
96 SG: Handelingen TK 22 February 1935.
97 SG: Handelingen TK 19 February 1936.
98 SG: Handelingen TK 15 February 1939. For other criticism of De Vries-Bruins: Handelingen TK 9 February 1937, 28 February 1938.
99 SG: Handelingen TK 19 February 1936.
100 SG: Tweede Kamer Memory of Answer Minister Colijn 1 June 1936.
101 CE Bijlagen 1937/1938, 44.
102 CE 1938/1939, p. 19.
103 CE Bijlagen 1937/1938, 44.4.
104 CE Bijlagen 1938/1939, 9.1.
105 CE 1938/1939, p. 20.
106 CE 1938/1939, pp. 16–21.

107 Ramsoedh, 'Suriname 1933–1944'; H. Ramsoedh, 'Rumcola en Yankeedollars', *OSO* 14 (1995), pp. 134–47; H. Buddingh', *De geschiedenis van Suriname* (Amsterdam: Nieuw Amsterdam, 2012), pp. 279–85; Hoefte, *Suriname*, pp. 93–5.
108 GS 1260.
109 GS 1260, 'Uit het verslag van den leproloog over 1944', pp. 5–6.
110 GS 1264, 'Verslag lepradienst over 1946'.
111 CE 1946/1947, Bijlagen 443/2.
112 J. Boom, 'Het levensverhaal van Hubert Willems. Opstand in de leprozerie Groot-Chatillon', *OSO* 22 (2003), pp. 117–22.
113 National Archive, The Hague, Governor of Suriname 1885–1951, 858, letter Jhr G. M. Versoyck, director Nederlandsche Rode Kruis, to the Governor, 12 August 1950.
114 P. J. van Eyk, 'Oorlogsjaren in Suriname. Nederlands koloniaal beleid binnen Amerikaanse marges', *OSO* 14 (1995), pp. 148–57.
115 D. van Hinte-Rustwijk and G. van Steenderen-Rustwijk, 'Van bedrijfsschade tot verzuilde paria', *OSO* 22 (2003), pp. 10–20, on pp. 18–19; M. Themen-Sliggers, 'De maatschappelijke aspecten van lepra', *OSO* 22 (2003), pp. 104–11.

8
Colonial medicine and folk beliefs in the modern era

In the early twentieth century, Dutch doctors and public health officials tried to come to grips with the Afro-Surinamese belief in treef and its influence on the execution of public health policies. Both the Afro-Surinamese and new British Indian and Javanese migrants maintained their beliefs and practices about leprosy. This on-going adherence to folk beliefs and practices alongside Western medical knowledge necessitated a response from Dutch colonial medicine. If modern leprosy politics were to succeed, some degree of cooperation and compliance from the population was necessary.

On the one hand, the Dutch considered folk practices as harmful. They were associated with the use of poison or black magic by the Afro-Surinamese in attacks on white rulers. However irrational these fears might have been, they were ingrained in the mindset of many Europeans during the age of slavery. On the other hand, from the early seventeenth century, Dutch adventurers operating in colonial medicine had been interested in studying African and Amerindian beliefs and practices about health and disease in the hope of gaining medical knowledge that would be useful for survival in tropical climates. Faced with a continued belief in treef and other folk beliefs among the large majority of the population, practitioners of Dutch colonial medicine investigated these local beliefs. In line with its own historical tradition, that of the adventurers of Dutch medicine overseas, Dutch colonial medicine in Suriname was primarily concerned with possible practical implications of folk beliefs and practices when translated within its own scientific framework. Folk beliefs were not seen as a possible alternative to Western science and medicine on a conceptual level.[1] Dutch colonial medicine found elements in folk beliefs useful for its own health propaganda and communication, while at the same time emphatically rejecting the folk medicine practitioners' world view underlying these beliefs. In this sense, Dutch colonial medicine did not limit itself to the interventions from above based on biomedical knowledge that historians have found typical of 'Imperial Tropical Medicine', but actively sought the compliance of the population.[2]

Fears of Afro-Surinamese beliefs and practices

In the era of modern leprosy politics, Dutch fears about leprosy remained focused on the Afro-Surinamese. The vast majority of leprosy sufferers continued to belong to this group, even though there was an increase in British Indian and (to a lesser degree) Javanese sufferers. In 1929, a survey of the Public Health Service concluded that of 916 leprosy sufferers registered by the authorities (407 of whom were admitted to the asylums) 723, or almost 80 per cent, were Afro-Surinamese. A total of 141 (approximately 15 per cent) of the sufferers were British Indian, and fifty-two or approximately 5 per cent were 'Dutch Indian' or Javanese. The survey did not mention European sufferers.[3] Dutch colonial medicine was therefore much less interested in the folk beliefs of the new immigrant groups (British Indians and Javanese) around leprosy than in those of the Afro-Surinamese.

Dutch physicians in Suriname were particularly afraid of the Afro-Surinamese as potential transmitters of the leprosy bacillus to other population groups. When active as 'peddlers and prostitutes' it was feared that the Afro-Surinamese would come into close physical contact with members of other ethnic groups.[4] Stereotypes of the role of sexuality in the transmission of leprosy, originating in the age of slavery, were still very much alive among Europeans in 1900 and later. For example, a Dutch journalist explained to his readers in the Netherlands that the Afro-Surinamese took utmost pleasure in idolatrous and lascivious dances that stimulated extramarital sexual contacts, and so contributed to the further spread of leprosy.[5]

Another problem for Dutch physicians was that the Afro-Surinamese continued to believe in the treef according to a 1928 Public Health Service survey. The survey conducted among leprosy sufferers, hospital patients and schoolchildren revealed that the treef belief did not necessarily stand in opposition to the view of leprosy as a contagious disease, since according to this belief a treef could be transmitted by contagion. However, for doctors, such as Lampe, the treef belief did prevent people in Suriname from viewing contagion as the *primary* transmission mechanism and to act accordingly.[6]

There was another subtext here. The Dutch perceived the continued Afro-Surinamese's conviction about their belief systems and healing practices as a possible undermining of order in the colony. These fears, originating in the age of slavery, continued to exist after emancipation. A glimpse of the threat of social disorder can be seen in Paramaribo on Wednesday 29 June 1883. The occasion was the funeral of a man called Samuel who had died as a patient in the Military Hospital. To the Afro-Surinamese, Samuel was known as 'tata' (father) Samweri. He was a wisiman, a practitioner of the black arts, and he used his magic for his clients' social and financial advancement. Some prominent citizens of Paramaribo had used Samweri's assistance to climb the social ladder. It was believed that Samweri was especially helpful to the young female mistresses of wealthy citizens by using his magic to bind their lovers to them, and it was this in particular that roused the wrath of the local Afro-Surinamese. Freed from their fear of Samweri by his death, the Afro-Surinamese rioted at his funeral and attacked the mistresses' dwellings. However, to the whites there was something even more sinister about the wisiman. Some people had been jealous of a respectable citizen in an honourable position, so they had employed Samweri to have him removed. Samweri gave the man leprosy using a method so horrible that the Surinamese newspapers did not divulge the details. It was feared that this victim had not been the only one who Samweri had afflicted with leprosy, or alternatively made lame or insane.[7]

Whether or not the story of Samweri giving leprosy to people by using his black arts is true cannot be ascertained. The story stands in a long tradition of fears among the Dutch rulers about being attacked by Africans with poison or other 'wisi' – black magical methods.[8] The story about Samweri points to the extent to which leprosy and its associations with Afro-Surinamese medical and magical practices (in this case performed by the wisiman) were disturbing to the Dutch. One solution was to prohibit these practices. In the early years of the twentieth century, a member of Dutch Parliament asked the government to end the 'crimes' of the obiah men who he said abused their influence over the population and did not hesitate to use poison and terrorism to perform their mischief. The police were afraid of the obiah men and a prominent socialist member even asked for a secret Suriname police to fight them.[9] However, prohibition was not sufficient to change the beliefs in the minds of the Afro-Surinamese.

From the perspective of the new modern leprosy politics, the essential problem was that some kind of patient cooperation and compliance with their doctors was necessary, especially if outpatient treatment was to be a successful alternative to segregation. In order to ensure this cooperation, some understanding of the patients' belief systems was necessary. Later, in 1959, the dermatologist Robert Simons (a trainee of the Medical School in Paramaribo) would complain that an interest in the patients' beliefs was singularly lacking in textbooks of tropical medicine. Simons believed that the 'reservoir' of the human being (his words) in which microbes and insects, symptomatology and therapy were important, was missing from Western medical professionals' understanding of diseases. Thus, Western professionals lacked an understanding of non-Western perspectives and beliefs. To Simons, the fight against tropical diseases including prevention could not be successful unless non-Western beliefs (what he still regarded as 'superstitions') and the role of the Western doctors' counterparts, the folk healers and priests, were studied.[10] Simons' viewpoint appeared to be quite revolutionary, but in his youth, in the 1910s and 1920s, it had been rather common among medical practitioners in Suriname. Although non-Western folk beliefs were not seen as a serious alternative to Western science, Dutch colonial medicine recognized that they exercised a great hold on the non-Western part of the Surinamese population. This in itself was sufficient reason to study these practices and beliefs and to devise ways to address them, or even make use of them in health propaganda if possible.

Treef and smallpox vaccination

When smallpox vaccination programmes were introduced in Suriname in 1900, the discussions about compulsory vaccination of schoolchildren showed how deep and sensitive beliefs and feelings about observing the treef were in Suriname, and how careful the colonial state was in addressing them. In the Netherlands, cowpox inoculation for the

Modern colonial medicine and folk beliefs 203

prevention of smallpox was obligatory for schoolchildren from 1823 to 1857. Since only a minority of children attended school at that time, the policy was of limited effectiveness. The catastrophic results of the smallpox epidemic of 1871–1873, with almost 16,000 deaths in 1871 alone, led to a new Law on Infectious Diseases (1872). Schoolchildren were once again obligated to show a certificate that they had been inoculated with the cowpox vaccine. By 1900, compulsory education was mandated by law for all Dutch children between the ages of seven and twelve. In principle, therefore, all children had to be inoculated including those with Christian parents who had moral and theological objections against vaccination.[11]

Small pox epidemics had ravaged Suriname in the eighteenth and early nineteenth century. In 1819 and 1820, 10,000 people (mostly slaves) had fallen victim to the disease. But since the last limited epidemic in 1823, the disease no longer posed as much of a danger in Suriname as it did in the Netherlands. In 1884, a vaccination station was established in Paramaribo for inoculation on a voluntary basis.[12] Nevertheless, in 1903, Governor Lely proposed to also introduce compulsory vaccination for schoolchildren in Suriname. Compulsory education of children between seven and twelve years of age had been introduced in Suriname as early as 1876, a quarter of a century before the Netherlands.[13] In 1903, the Surinamese school population numbered approximately 6,300 children. However, the colonial government was alert to the problem of possible 'conscientious objectors' to compulsory vaccination. This included some Christians, but in the Surinamese context, more important were the objections of parents whose children had a cow as their treef. To these parents it was unthinkable that their children should be inoculated with a cowpox vaccine since it violated their taboo animal, which risked giving their child a disease or worse. High-ranking civil servants of the colonial administration, such as Herman Benjamins, the inspector of education, were well aware of this problem. They estimated that resistance against vaccination among this group would be so significant that it would endanger the whole educational system if vaccination was enforced. At the same time, Benjamins did not wish to give legitimacy to the belief in the treef by officially recognizing it as a valid ground to refuse vaccination.

The potential number of objectors was high owing to the fact that belief in the treef was widespread. A 1908 school survey showed that 75 per cent of all schoolchildren had a treef.[14] Of even more concern was that a belief in the treef was not limited to the Surinamese lower classes, but rather it was also spread among the middle and upper classes, which made it paramount for the government to proceed carefully. In December 1903, the validity of the treef belief was an important critique for the opponents of the new vaccination edict in the Colonial Estates. Members of the Estates expressed resentment at calling the treef belief a superstition, and claimed that all scientific beliefs had once begun as superstitions. One member suggested giving the treef another more scientific name – idiosyncrasy, a contemporary name for a hypersensitivity to foodstuffs for example. He proposed to give people with idiosyncrasy the right to refuse vaccination.

The government had an alternative strategy. Without denying that there were scientific grounds for the belief in the treef or expressly referring to treef in the text of the edict, school directors of private schools were given permission to ask for vaccination exemptions for their pupils. This did not satisfy all the members of the Estates. Most schoolchildren, approximately 4,000, went to the fee-paying private Moravian and Catholic schools, whose directors could ask for exemptions. The rest, approximately 2,300, from the poorer segments of society went to the free public school and would have to be vaccinated. However, ultimately the Colonial Estates accepted the new regulation of a compulsory vaccination with an escape clause for members of the wealthier classes.[15]

Thinking about the treef

The Surinamese colonial government and medical practitioners felt the need for compromise to obtain compliance with their vaccination programme. However, they did not wish to give the impression that they had given in to African superstitions. By suggesting a possible scientific ground for the treef belief, they evaded that pitfall. The key word here was idiosyncrasy. The notion of idiosyncrasy had long-time roots in Western medicine. From the perspective of classical medicine, each individual had their own unique combination of elements, such as bodily fluids, and therefore his or her own unique sensitivities and reactions. In nineteenth-century Western medicine, this idea, though modified, survived and was used to explain the diverse reactions of individual patients to medicines.[16] The concept of idiosyncrasy gave doctors in Suriname a sort of theoretical handhold on the treef belief. They could take belief in the treef seriously by translating it into their own Western concept without being thought to be superstitious. For instance, Wilhelm Essed, another graduate from the Medical School in Paramaribo (and in the 1920s and 1930s a prominent parasitologist in the Dutch East Indies), thought that some of the people who had a treef (perhaps one in twenty) actually suffered from an idiosyncrasy.[17] The *Encyclopedia of the Dutch West Indies*, published during the First World War and co-edited by Benjamins, stated that an idiosyncrasy for a specific food or for spices, medicines or other items that resulted in more or less serious and lasting skin diseases was well known in Suriname. This kind of idiosyncrasy could be inherited. But having a treef was not always regarded in the same way as having an idiosyncrasy. Benjamins explained that in Europe the existence of an idiosyncrasy was established by scientific research and medical investigation, but in Suriname the existence of a treef was established by all kinds of supernatural means that he clearly considered to be heathen mumbo-jumbo. Still, Benjamins could not just shrug his shoulders about this belief since most of the Afro-Surinamese thought that the treef was the only cause of leprosy. This changed matters. The colonial government and colonial medicine had to manage this superstition and the *Encyclopedia* advocated more research on treef beliefs among schoolchildren.[18]

Opinions on the treef belief among Dutch medical practitioners and intellectuals in Suriname varied between those who saw it as an outright superstition, as presented in a leading Dutch liberal daily in 1923, and as a danger to leprosy prevention politics as believed by Flu and Benjamins, and the belief held by some doctors, for example, T. May, who saw the treef belief as based on observation and thought it held a kernel of truth.[19] Whatever opinions on the treef belief were, the health service had to address it.

The treef survey

To the architects of the edict of 1929, compulsory segregation was self-defeating unless the majority of the population supported it. Segregation had to be made acceptable by the prospect of medical treatment. To gain further support from the Surinamese people, Lampe felt that their beliefs had to be taken seriously. If the Afro-Surinamese felt that Western practitioners looked down upon their beliefs, they would be less inclined to cooperate and inform the authorities about possible cases of leprosy. That this cooperation was needed and that it could be organized became evident when Lampe and his collaborators tried to ascertain the exact number of leprosy sufferers in 1927. Previously, attempts to trace persons who had been declared suspect by the leprosy committee over the years had failed. Neither the registry office nor the immigration department had any knowledge about the whereabouts of the suspects. An advertisement in local newspapers asking for their whereabouts had some success. A number of people were reported by their neighbours as suspected of leprosy, and some suspects asked for a doctor's visit and turned out to be sufferers. By December 1927, eleven suspects were visited at home twice a week and four patients in an early phase of the disease were in treatment at the then newly established outpatient clinic. The new patients spread the news and more new patients followed. In less than a month, a doctor (Ch. Simons) had to be specially charged with the treatment of the sufferers and by the end of 1928, the doctors were confident that all non-segregated sufferers in Paramaribo, approximately 300 in total, had been found. It was believed that only a small number of suspects were not identified and they were thought to be deceased.[20]

This 'success' showed that the population's cooperation with the execution of leprosy politics could be organized. At the same time, it was a long-standing Dutch idea that belief in the treef was a hindrance leading to fatalism and non-cooperation. Was this necessarily true? The reactions to the advertisement suggested otherwise. Lampe felt it was time for a thorough survey of both the qualitative content and the quantitative dissemination of the treef belief among the Surinamese.

Lampe organized a survey among hundreds of people. These included up to 400 leprosy patients (in particular Afro-Surinamese) in treatment with the health service or admitted to the Majella asylum, over 200 Surinamese-born patients with other diseases who were in a hospital on the day of investigation, and schoolchildren in a number of private and public schools. The first two groups were directly asked about their treef. The children had to write an essay on the subject.[21] Of the 400 sufferers, 348 had a treef (87 per cent). A total of 75 per cent of the sufferers were convinced that leprosy could only develop if the treef was violated. Almost half of the sufferers (47 per cent) believed that the disease had developed because they had (repeatedly) violated their treef, but others thought that this violation had only prepared the ground for transmission of the disease by contagion (22 per cent), and a few (2 per cent) thought that the violation had prepared the ground for inheritance of the disease. To Lampe this was an important research finding, since it showed that a belief in the treef was not necessarily opposed to a belief in the scientific concept of contagion, or even to belief in inheritance. Congruence between science and folk beliefs was possible. On the other hand, only 15 per cent of the interviewed sufferers did not believe in the treef, 10 per cent were uncertain of the cause of leprosy, and a mere 4 per cent believed only in contagion. Therefore, the treef belief was not in itself in opposition to a belief in contagion, but it was an obstacle to people believing *only* in contagion.[22]

More than half a century of compulsory Dutch school education in Suriname had not done away with magical folk beliefs. Lampe classified beliefs in various kinds of treef. People could have a hypersensitivity or idiosyncrasy for a variety of kinds of food or stimulants, causing skin afflictions, gastric disturbances, nervous diseases, and other health problems, and specifically in leprosy sufferers causing increases in signs of the disease and fever. When they attributed this to a treef, Lampe called this the *experience* treef. People could have an *inherited* treef from their father, and would not use this treef. Some thought that an experience treef could change every seven years, but an inherited treef could never change. A *dream* treef was revealed before one's birth in the dreams of a family member or seer. One could establish one's treef after showing symptoms of psoriasis (treefvlekken) or leprosy by studying the resemblance of the skin spots with the outline of an animal, or by a child's dislike of a particular food. One could also believe in a non-individual, *general* treef because of the properties of a certain kind of food. For instance, the 'heating properties' of pig's meat meant that pig became a general treef for everyone. The treef could also be other food including the following: mammals, reptiles, bird, fish, vegetables, fruit, herbs or cassava. One could also have several treef at the same time. Of the 348 sufferers with a treef, the most common treefs were a cow (35 per cent) or pig (33 per cent).[23]

Lampe's survey revealed how influential the treef belief was in patients' health behaviours. Leprosy sufferers often experimented with a variety of food to ascertain the effects of a treef (the *experimental* treef), an activity that could lead to an almost complete refusal of food.

Patients asked their doctors which medicine belonged to which treef. Patients then used the powder of their treef to dress signs of leprosy on the body. In early stages of the disease they went to treef experts ('treefkenners'), obiah men, lukumen, or even black magicians (wisimen) to ask for a medicine on the basis of their treef. Many believed that a treef could disappear when one got older or with the help of the rituals of the lukuman or the wisiman.[24] When one entered a traditional household in Suriname, the guest was asked about his or her treef so the host would not offer the wrong food to the guest.[25] Cultures could clash about the observance of the treef in interracial households. One leprosy patient, a girl of eighteen years of age, had a Dutch father and a coloured mother. When she was a young child, she dreamt of two Indians who told her that her treef was cow. However, her father did not believe in treef and forced the girl to eat beef after the death of her mother. One year later she became very ill, and she claimed that her leprosy spots became larger when she ate her treef.[26]

Among children, the treef belief showed no sign of disappearance. Of all the children in fee-paying schools, 58 per cent wrote that they had a treef, as did 71 per cent of the children in the free public schools. However, these results were suspect and the percentages were probably higher, especially in the fee-paying schools. The children would hide their treef beliefs when they had the impression that they would be ridiculed by their Dutch doctor. In one public school, over 90 per cent of the children reported a treef, and Lampe thought that this result was closer to the actual percentage for the whole Surinamese school population.[27]

Lampe realized that when dealing with leprosy in Suriname, public health officials had to take the impact of the treef belief into account. He thought a belief in treef was more than mere superstition, and even saw a scientific foundation. After all, contagion with the leprosy bacillus did not automatically lead to leprosy, but 'something else' had to come in as well. For the Surinamese population, the treef was an explanation of this 'something else'. Patients with leprosy, folk medical practitioners, and Afro-Surinamese nurses believed in the treef and observed its effects when a sufferer's condition deteriorated following the consumption of certain foodstuffs. Therefore, Lampe advocated more research. However, he also warned about the dangers of the treef belief provoking a disregard for the risks of contagion, a lack of trust in the doctors resulting in delayed diagnosis and treatment, and a reckless observance of the treef resulting in an insufficient diet and a weakened constitution.[28]

Although the treef belief had to be investigated, the healing methods of Afro-Surinamese practitioners by definition did not fit in the world view of Dutch colonial medicine and were still considered to be dangerous. To the Dutch, these methods had no place in a modern colonial state. Therefore, incorporating beliefs associated with the treef into medical health propaganda also served another goal of the Dutch – to undermine the hold of Afro-Surinamese medical practices on the colony population. Rather than make a frontline attack on belief in treef, doctors had to tread carefully when managing sufferers and patients with those beliefs, and not ridicule them, but rather emphasize the possible overlap with notions of contagion. This was not a unique approach to leprosy: the Colonial Report of 1937 concluded that when managing tuberculosis, a belief in treef did not need to clash with a belief in infection.[29]

A belief in treef did not mean that the Afro-Surinamese were not interested in European treatment methods. Anthropological research in Paramaribo in the 1930s discovered that although the lukuman was the practitioner to consult when a disease could have been caused by a violation of the treef, the Surinamese readily made use of other practitioners, whether the obiahman, the wisiman, or the white doctor.[30] The Djuka, one of the most important Maroon tribes in the interior, did not believe in natural causes of disease (rather than supernatural), but nevertheless were happy to use European medicines.[31] Patients were nothing if not pragmatic and opportunistic, regardless of their underlying world views; a medical pluralism that is not only characteristic of multicultural societies, but has been documented for the ethnically and culturally homogeneous Netherlands in this period as well.[32]

Vampires, insects, and leprosy

A belief in the treef was also of interest to Western scientists and medical practitioners because not all of the mysteries of leprosy had been solved. The general consensus was that the disease was contagious, and caused by the Hansen bacillus. However, by 1950, the Hansen bacillus had still not been cultivated in vitro, and in some manifestations of the disease there were few, if any bacilli present. '*Mycobacterium leprae* ... has not come up to expectations', concluded Lampe in 1949.[33] Leprosy continued to be a mysterious disease. The disease did not develop in every person to whom the bacillus was transmitted. Did a weakened constitution, perhaps owing to diet for which one had an idiosyncrasy, play a role? How was the bacillus transmitted? These were important questions throughout the Dutch colonial rule in Suriname, and they have still not been conclusively answered in the early twenty-first century. Could folk beliefs, superstitious as they were, offer clues to answers?

In the 1930s, Lampe (who had by then moved to the Dutch East Indies and was working as director of the institute of leprosy research in the capital Batavia, now Jakarta), sponsored research into the endemiology (study of endemic disease) of leprosy in order to gain a greater understanding of leprosy's transmission mechanisms. During the Second World War, Lampe was interned in a Japanese prisoner-of-war camp on Java, and he spent part of his time working on the results of his research in 'leper' and 'non-leper households' in the city of Batavia. His findings suggested that the endemic dissemination of leprosy was dependent on the extent of social intercourse between carriers of the disease and others. Bed contact was of statistical significance here.[34] This could explain why leprosy was more common among the poor who had limited living space and shared beds. However, the question still remained how the transmission mechanism exactly worked.

Folk beliefs might give Western medicine clues to further investigation. In Suriname, folk beliefs were not limited to the treef. People, who were often bitten at home by bedbugs or rats at night were suspected by their neighbours of having leprosy. Could this offer a clue? May had followed this clue in a publication in 1928. As Leonard Rogers supposed, if the leprosy bacillus was transmitted through the skin, the vectors could be bloodsucking insects. This would explain why leprosy was more prevalent in the tropics where insects could flourish for a short time during high temperatures and heavy rainfalls, and why leprosy was a social disease and a disease of poverty that disappeared in conditions of better hygienic.[35] After the Second World War, Robert Simons (then practising in Amsterdam) followed up Lampe's research. Simons was requested to write a report on leprosy for the colonial government in Suriname. He visited the colony and described distinct focal points of contagion. A domestic partition-wall theory explained that districts and streets could be focus points where insects in beds and furniture might transmit the bacillus. The theory explained why people with a lower social-economic status, who were also vulnerable because of their poor diet and constitution, were over-represented among leprosy sufferers. The bedbug was Simons' primary suspect, but at an international dermatology congress in Brussels in 1948, the cockroach was proposed as a more likely candidate. Although the domestic partition-wall theory was unproven and not generally accepted, it shows that Dutch doctors paid close attention to folk beliefs on leprosy.[36]

In their writings on other folk beliefs, May and Simons referred to a belief in the demons Azema, a blood-sucking vampire (azeman), and the fio-fio, who manifests himself as a bedbug. Fio-fio is a form of bewitchment. Both demons come out at night to suck the blood of their victims. The fio-fio is family of the spider god, Anansi, the jester among the Surinamese gods and demons, and the itch caused by his bite is the god entering into the body. When disease follows, this is a divine judgement, unless a demon (and not a god) is responsible. In the latter case, magical treatment can arrest the disease.[37] As in the case of the treef, the belief in the demonic bedbug was not incompatible with the practices of Western science, although it could interfere with early diagnosis and treatment.

The British Indians

Dutch colonial medicine in Suriname was much less interested in the folk beliefs of the new immigrant groups in the population of leprosy sufferers (British Indians and Javanese) because the large majority of all sufferers were of African descent. Nevertheless, like the Afro-Surinamese, the British Indians had their own traditions of folk beliefs around leprosy that influenced their health behaviour around leprosy.

Initially, they were regarded by the Dutch as especially at risk for contracting the disease because of their supposedly unhygienic customs, cramped living conditions, and unhealthy diet. In 1902, sixty-four of the 149 patients in the Groot-Chatillon government asylum were British Indians.[38] One solution to manage leprosy under the indentured labourers was to repatriate this group. In 1905, thirty were returned to India, and in 1911, another twenty-eight. In 1912, twenty-nine of 106 patients in Groot-Chatillon were British Indians.[39] They continued to be an important and often disturbing presence in the asylum, as will be described in the next chapter.

The Hindu culture's attitude towards leprosy in British India shows similarities to the Afro-Surinamese. Historian Jane Buckingham describes how in the Hindu tradition the disease is not simply seen as a physical sickness but rather as the manifestation of a spiritual condition. For Buckingham, 'The Hindu sāstric tradition identified leprosy as a condition entailing profound ritual pollution ... Sāstric authorities allowed for the outcasting of leprosy sufferers displaying ulceration.'[40] The outcasts lost their ritual and social identity, though their families were still expected to take care of them. However, sāstric law primarily applied only to the highest-ranking Brahmin or priest caste among the Hindus and was less relevant for (or even ignored by) other castes. According to Buckingham, in Muslim law the concept of outcasts was unknown. Certainly, not all leprosy sufferers were outcasts in British India, and there was a high degree of physical interaction between sufferers and others, especially in business dealings, which was a situation again not unlike the experience of the Afro-Surinamese. Stigmatization, avoidance, and abhorrence were primarily directed at beggars and vagrants with leprosy.[41]

There is some evidence of a cultural exchange of views on the relationship between food and leprosy between the Afro-Surinamese and the British Indians. One British Indian leprosy sufferer remembered the advice of family, friends, and family doctors to avoid spicy food as well as fish, duck, and chicken butchered from the market.[42] Another British Indian sufferer who had converted to Islam said that Muslims had a treef, but Hindus did not. He mentioned a relationship between treef and leprosy and black pepper and unscaled fish.[43] A third sufferer, admitted to the Bethesda asylum in 1938 at the age of seven, remembered how a British Indian man accused a woman of having bewitched another woman with a 'winti', a spiritual wind. The bewitched woman contracted leprosy.[44]

Like the Afro-Surinamese, the British Indians carried their own plant cures with them to the Americas. Mr. S. B. was born in 1933 and diagnosed with leprosy around 1940. Although British Indian by birth, he was a Catholic by religion, and so he was sent to the Catholic leprosy asylum Majella (or he converted as a condition for admittance to avoid the government asylum). In Majella, he managed to evade taking his medicine, but his mother gave him two bottles of neem extract each week. Neem (Azadirachta indica) was originally an Asian tree, but it now grows throughout the tropics and often in the gardens of Hindustani in Suriname. Taken as a tea or in a decoction, neem is considered by many British Indians as a miracle cure for all kinds of diseases, and for purifying the body.[45]

From the British Indian perspective compulsory segregation of their leprosy sufferers may have been considered controversial, since the sufferers would not necessarily have been outcast by their own communities and families, and plant cures were thought available. This might have caused further resentment among the British Indian sufferers in the Groot-Chatillon government asylum.

The Javanese

The third largest group among asylum patients consisted of the Javanese. This group had its own folk beliefs around leprosy as well. However, although in 1907 and 1909, medical investigations of Javanese migrants in Suriname discovered a high prevalence of other skin diseases, leprosy was relatively rare. In 1909, a report mentioned only four Javanese patients in Groot-Chatillon, while the yaws station in the village of Groningen had predominantly Javanese patients.[46] This suggests that the Javanese migrants to Suriname came from island regions that were not pockets of leprosy. In the history of leprosy in colonial Suriname, Javanese leprosy sufferers were not so visible, but they did exist. After the inauguration of the new 1929 leprosy politics, most of the Javanese asylum patients (thirty) were repatriated to the Dutch East Indies.[47]

For the Javanese, as for the Afro-Surinamese and British Indians, cultural concepts of disease were suffused with spiritual meanings. Animistic beliefs continued to hold an important role in managing health and disease on Java. Dutch colonial observers in the Dutch East Indies and modern-day anthropologists and sociologists have described how the Javanese believe that diseases can be caused by worms, poison, magic, or (unhealthy) winds. Angry spirits can bring disease, and so can a transgression of customs and law and inappropriate social and religious behaviour.[48] A 1964 study on the Javanese in Suriname concluded that they believed both in the Koran and spirits. In everyday life, the spirits had the greatest influence. Disease could be a consequence of black magic and was fought with other magic, such as wearing amulets. Women were especially skilled in magic, including the use of medicinal herbs and casting spells, but the specialist was the male dukun.[49] The dukun used combinations of herbal medicine and magic in curing his patients. One rationale for these treatment methods was the idea of a transmigration of properties from the remedy to the patient. In the case of leprosy, a Dutch study in the East Indies from the 1890s listed the use of cicak, a small lizard (Hemidactylus frenatus), because a lizard can grow back a lost tail.[50] Another Dutch study in 1918 concluded that the Javanese had no medicine against 'Barah' (leprosy without swellings and ulcers) or 'Boeroeg' (leprosy with swellings and ulcers).[51] Studies of Javanese folk beliefs were never undertaken in Suriname under Dutch rule. The relative number of Javanese in the total of leprosy sufferers was too insignificant for colonial medicine for serious research to be undertaken on their beliefs and practices around leprosy.

Conclusion

As far as can be ascertained, in the modern colonial state non-Western cultures held on to their own beliefs on leprosy's aetiology. Although pragmatically willing to use Western medical treatment when seen as advantageous, leprosy was framed within perspectives of spiritual forces, pollution, and taboo violation. Notwithstanding decades of compulsory Dutch education, the large majority of leprosy sufferers and their families continued to believe in the role of the treef. This does not necessarily mean that they did not understand Western ideas about transmission and contagion. Modern studies in the late twentieth century of patient–doctor interaction show that acceptance of the doctor's views is primarily an understanding of whether they fit with the patient's cultural beliefs. As researcher Jenny Donovan concluded, 'Patients do not just accept what they are told, it has to make sense and be justifiable within their ways of looking at the world.'[52]

To Dutch colonial medicine, the Afro-Surinamese and other groups' adherence to cultural beliefs endangered early leprosy detection and prevention policies. In order to ensure greater compliance with the reformed leprosy politics, the Dutch developed a strategy in which the treef belief was not simply dismissed as a superstition, but rather taken seriously as a folk intuition of a hypothesis with a scientific basis – the role of idiosyncrasies. In line with its own historical tradition of Dutch medical adventurers overseas, Dutch colonial medicine in Suriname was primarily concerned with the possible practical implications of folk beliefs and practices as translated within its own scientific framework. Folk beliefs were not seen as a possible alternative to Western science and medicine on a conceptual level.[53]

While medical pluralism, the use of both European and complementary non-European medicine and treatment, characterized the different ethnic groups in colonial Suriname, Dutch colonial medicine sought to incorporate folk beliefs in a more open-minded health communication directed at leprosy sufferers and their families. This and the emphasis on medical treatment rather than isolation were the 'carrots' offered by Dutch colonial medicine, without throwing away the 'stick' of compulsory segregation. Implementing strategies to ensure compliance with leprosy policies never did affect the essential authoritarian character of these policies.

Notes

1 S. Snelders, *Vrijbuiters van de heelkunde. Op zoek naar medische kennis in de tropen 1600–1800* (Amsterdam: Atlas, 2012).
2 On the concept of 'Imperial Tropical Medicine': M. Worboys, 'The emergence of tropical medicine: A study in the establishment of a scientific specialty', in G. Lemaine, R. Macleod, M. Mulkay and P. Weingart (eds.), *Perspectives on the Emergence of Scientific Disciplines* (The Hague: Mouton, 1976), pp. 75–98; J. Farley, *Bilharzia: A History of Imperial Tropical Medicine* (Cambridge: Cambridge University Press, 1991); D. M. Haynes, *Imperial Medicine: Patrick Manson and the Conquest of Tropical Disease* (Philadelphia, PA: University of Pennsylvania Press, 2001).
3 There were a further forty-four persons registered as suspect, but they were not differentiated according to race: P. H. J. Lampe and C. Simons, 'Lepra in Suriname', *Nederlandsch Tijdschrift voor Geneeskunde* 73 (1929), pp. 4903–15, on pp. 4905, 4912. See also League of Nations Leprosy Committee, League of Nations Room, United Nations, Geneva, 'Suriname' (Courtesy of Jo Robertson).
4 Salomons in Handelingen Tweede Kamer 19 December 1896, *Handelingen Staten-Generaal*: www.statengeneraaldigitaal.nl/. Twenty years later Wolff was still in fear about the consequences of sexual contacts between white

and non-white: J. W. Wolff, 'Het lepra-probleem in Suriname', *Geneeskundig Tijdschrift voor Nederlandsch-Indië* 65 (1925), pp. 572–87.
5   A. C. Wesenhagen, *Suriname, Iets over land en volk* (Amsterdam: J.H. de Bussy, 1896), p. 26. This book by a Dutchman born in Suriname is full of racist prejudices against Afro-Surinamese, Maroons, and Chinese.
6   P. H. J. Lampe, 'Het Surinaamsche treefgeloof. Een volksgeloof betreffende het ontstaan van de melaatschheid', *West-Indische Gids* 10 (1929), pp. 545–68, on pp. 545–6.
7   *Suriname, Koloniaal Nieuws- en Advertentieblad* 1 June 1883, nr. 46, n.p.; *De West-Indiër* 6 June 1883, nr. 45, n.p., www.delpher.nl/nl/kranten/ (hereafter SN).
8   Snelders, *Vrijbuiters van de heelkunde*, pp. 181, 205.
9   J. M. Plante Fébure, *West-Indië in het parlement 1897–1917, Bijdrage tot Nederland's koloniaal-politieke geschiedenis* ('s-Gravenhage, Martinus Nijhoff, 1918), p. 84.
10 R. D. G. P. Simons, *Bijgeloof en lepra in de Atlantische Negerzônes* (Paramaribo: Radhakishun, 1959), pp. 5–6, 10.
11 W. F. H. Veldhuyzen, *Honderd en vijftig jaar pokken preventie* (Amsterdam: Scheltema and Holkema, 1957).
12 H. D. Benjamins and J. F. Snelleman (eds.), *Encyclopaedie van Nederlandsch West-Indië* ('s-Gravenhage: Nijhoff, 1914–1917), p. 282.
13 *Gouvernementsblad van de Kolonie Suriname* (hereafter GB) 1876, nr. 10.
14 H. D. Benjamins, 'Treef', in Benjamins and Snelleman (eds.), *Encyclopaedie*, pp. 685–7, p. 686.
15 National Archive, The Hague, Handelingen en Bijlagen van de (Koloniale) Staten van Suriname 1903/1904, session of 8 December 1903, pp. 73–81; session 13 December 1903, p. 85.
16 L. C. E. E. Fock, *Natuur- en geneeskundig etymologisch woordenboek* (n.p.: J. Noorduyn, 1855), p. 457.
17 W. F. R. Essed, 'Eenige opmerkingen naar aanleiding van de artikelen over treef en lepra in dit tijdschrift verschenen', *West-Indische Gids* 12 (1931), pp. 257–67.
18 Benjamins, 'Treef', in Benjamins and Snelleman (eds.), *Encyclopaedie*, pp. 685–6.
19 The treef as superstition: M. van Blankensteijn, *Suriname* (Rotterdam: Nigh and Van Ditmar, 1923), pp. 78–9; L. C. van Panhuys, 'About the "trefe" superstition in the colony of Surinam', *Janus* 28 (1924), pp. 364–8. Danger of treef belief for healthcare: P. C. Flu, 'Het een en ander over de besmetting met lepra', *Stemmen uit Bethesda* 29 (1924), pp. 56–7;

Benjamins, 'Treef', in Benjamins and Snelleman (eds.), *Encyclopaedie*, p. 686. More positive: SN: Baeza in *Onze West* 2 March 1904; T. May, 'Bijgeloof', in *Suriname* 6 April 1923.
20 Colonial Reports (hereafter CR) 1928; Lampe and Simons, 'Lepra in Suriname', pp. 4903–15, on p. 4904; G.-J. Hallewas, 'De gezondheidszorg in Suriname' (PhD thesis, Groningen University, 1981), p. 43.
21 Lampe, 'Het Surinaamsche treefgeloof', p. 551.
22 Lampe, 'Het Surinaamsche treefgeloof', p. 546.
23 Lampe, 'Het Surinaamsche treefgeloof', pp. 553–5.
24 Lampe, 'Het Surinaamsche treefgeloof', pp. 547–50. On the dressing of leprosy spots with treef powder: Benjamins, 'Treef', p. 686. On winti rituals and lukuman treatment methods: van Panhuys, 'About the "trefe" superstition', pp. 358–61; P. C. Flu, *Verslag van een studiereis naar Suriname (Nederlandsch Guyana.) Sept. 1927 – Dec. 1927* (Utrecht: Kemink, 1928), p. 83.
25 Lampe, 'Het Surinaamsche treefgeloof', p. 558.
26 Lampe, 'Het Surinaamsche treefgeloof', pp. 555–6.
27 Lampe, 'Het Surinaamsche treefgeloof', p. 551.
28 Lampe, 'Het Surinaamsche treefgeloof', pp. 566–8.
29 CR 1937.
30 M. J. and F. S. Herskovits, *Suriname Folk-lore* (New York: Columbia University Press, 1936), pp. 59, 103.
31 W. F. van Lier, 'Aanteekeningen over het geestelijk leven en de samenleving der Djoeka's (Aukaner Boschnegers) in Suriname', *Bijdragen tot de Taal-, land- en Volkenkunde* 99 (1940), pp. 130–294, on pp. 273–6.
32 S. Snelders and F. J. Meijman, *De mondige patient. Historische kijk op een mythe* (Amsterdam: Bert Bakker, 2009).
33 P. H. J. Lampe and R. Boenjamin, 'Social intercourse with lepers and the subsequent development of manifest leprosy', *Documenta Neerlandica et Indonesica de morbis tropicis* 1 (1949), pp. 289–346, on p. 290.
34 Lampe and Boenjamin, 'Social intercourse'.
35 May, 'Lepra', pp. 551–6. See also Benjamins, 'Treef en lepra', p. 215.
36 Simons, *Lepra: De lepra-bestrijding in Suriname*, pp. 3–7. See also R. D. G. P. Simons, *Lepra. De maligne contagieuze Morbus Hansen en de benigne niet-contagieuze Hanseniden* (Amsterdam: Van Holkema and Warendorf, 1948), pp. 15–17.
37 T. May, 'De lepra, haar voorkomen, verspreiding en bestrijding, in 't bijzonder in Suriname', I, *West-Indische Gids* 8 (1927), pp. 547–56, on p. 556; Simons, *Bijgeloof en lepta*, pp. 30–3. On Anansi: L. Lichtveld, 'Op zoek naar de spin', *West-Indische Gids* 12 (1931), pp. 209–30.

38 CR 1903.
39 CR.
40 J. Buckingham, *Leprosy in Colonial South India: Medicine and Confinement* (Basingstoke: Palgrave, 2002), p. 31.
41 Buckingham, *Leprosy in Colonial South India*, pp. 31–5; Jane Buckingham, 'The "morbid mark": The place of the leprosy sufferer in nineteenth century Hindu law', *South Asia: Journal of South Asian Studies* 20 (1997), pp. 57–80.
42 Interview by Melinda Reyme with Mrs D. M., Paramaribo transcript March 2014, 'Give Ex-Leprosy Patients a Voice' (unpublished manuscript).
43 Interview by Melinda Reyme with Mr S. H, Paramaribo transcript April 2014, 'Give Ex-Leprosy Patients a Voice'. Another Hindu informer, Mrs S. K., did not believe in the treef (interview 1 August 2013). Mr S. B., who was diagnosed with leprosy around 1940, also mentions a belief in the relation between leprosy and unscaled fish.
44 Interview by Melinda Reyme with Mrs E. S., transcript December 2013, 'Give Ex-Leprosy Patients a Voice'.
45 Interview by Melinda Reyme with S. B., 'Give Ex-Leprosy Patients a Voice'; Van Andel and Ruysschaert, *Medicinale en religieuze planten*, pp. 332–3.
46 H. L. C. B. van Vleuten, 'Rapport omtrent den toestand der Javanen in Suriname' in Hoefte, *Betovering*, pp. 63–7.
47 CR 1933.
48 L. Hesselink, *Healers on the Colonial Market: Native Doctors and Midwives in the Dutch East Indies* (Leiden: KITLV Press, 2011), pp. 9–11.
49 De Waal Malefijt, 'Javanese population', pp. 196–223. De Waal Malefijt' research has been criticized by Superlan, 'Javanese'. On the dukun in Javanese culture J. A. van Dissel, 'Iets over de godsdienst der Javanen', *Tijdschrift voor Nederlandsch-Indië* 3rd series, 3 (1869), pp. 375–85, on pp. 375–6. Gender dukun: Hesselink, *Healers*, p. 14. On traditional medicine: P. J. Veth, *Java. Geographisch, ethnologisch, historisch* 2nd rev. edn, vol. 4 (Haarlem: De Erven Bohn, 1907), pp. 275–80.
50 Hesselink, *Healers*, p. 18.
51 L. T. Maijer, *De Javaan als doekoen. Een ethnografische bijdrage* (Weltevreden: G. Wolff, 1918), pp. 40, 43.
52 Cited in Snelders and Meijman, *De mondige patien.*, p. 87.
53 Snelders, *Vrijbuiters van de heelkunde*.

9

Complex microcosms: asylums and treatments, 1900–1950

By 1950, a new kind of leprosy asylum had entered the Western public imagination. No longer a place of horror, the modern leprosy asylum was a key example of the benefits of the work of colonial medicine and religious missions. It was a place of orderliness and cleanliness where sufferers could lead a meaningful existence and receive medical treatment with the prospect of returning to society. However, this public image was contested by a revisionist historiography in the 1990s and early 2000s.[1] Here the modern asylum was portrayed as a key instrument of colonial rulers' attempts to turn colonial subjects into modern citizens. In the last decade, this image changed again. Historians have developed more nuanced images, and thus they highlight asylums as complex microcosms where the sufferers had agency and everyday life was continuously renegotiated.[2]

This chapter explores these complex microcosms of the modern leprosy asylums in Suriname. In the decidedly unmodern Batavia asylum of the nineteenth century, Christian missionaries had already been involved in inmate discipline in accordance with European notions of correct behaviour and customs. These missionaries gave leprosy care a central place in their missionary activities and the presentation of these activities to their co-religionists and financiers in Europe. This approach continued in the modern Catholic and Protestant asylums of Majella and Bethesda, although in another form and with increased participation and intervention by colonial medicine. Christian asylums were not isolated entities, however. Together with the Groot-Chatillon state asylum, Christian asylums were interconnected parts of a system of leprosy care that was created after accommodation between the colonial state and the Christian churches in the 1890s. What resulted was a system of care and treatment that ideally would return cured and grateful citizens back to society.

The aims and activities of the colonial state and the Christian religions went hand in hand. Although there were certainly friction and turf wars between the colonial physicians and Christian missionaries, their fundamental division of labour was never questioned. The asylums were for those sufferers whose families could not or wished not to keep them segregated at home. The Catholics and the Protestants took care of their co-religionists who were subjected to strict religious discipline and forbidden to marry. Those who did not wish to adhere to this discipline (often Afro-Surinamese, Hindu or Muslim male sufferers from lower social classes) and married patients went to the state asylum supervised by the colonial state. Unruly patients from other asylums were also sent to the state asylum. Here, Catholic and Protestant missionaries were present, but adherence to religious discipline was on a voluntary basis. Looking from 'above', the system was flexible enough to allocate a place to each sufferer. Looking from 'below', this situation can be questioned since the asylums were characterized by their own infrapolitics of friction and resentment. The permeability of asylum boundaries characterized by the movement of patients between asylums and the outside world, and even between asylums, was apparent. In everyday life, there were limits to the disciplinary power of the regimes in the asylums.

Historiography of the modern asylum

When the British traveller and writer Patrick Leigh Fermor visited the leprosy asylum on Chacachacare island off the north-western coast of Trinidad shortly after the Second World War, he was much impressed by its humanitarian achievements. In his 1950 book, *The Traveller's Tree*, Fermor wrote about his visit.

> Nothing, at first, distinguished the lepers of Chacachacare from the other inhabitants of the West Indies. Some were working under the trees, and others were talking and sitting in the sun on the balconies of their little cabins … The ones whom we saw eating in their little dining-room, chattering and laughing and interrupting each other, or sitting on a bench and talking to one of the nuns, appeared as happy and as normal as though they did not know that leprosy existed.[3]

Fermor painted a picture of staggering contrast between this new and modern leprosy asylum and the debasing and stigmatizing treatment of leprosy sufferers of old. For Fermor,

> One realizes what a gulf now exists between the bells and the clappers, the cries of Unclean, the outlawry, the official, and sometimes religiously solemnized, death-in-life of the lepers of the past. [And though segregation of the sufferers is still painful] once they are installed in a place like Chacachacare, all is done that ingenuity and humanity can invent to make their lives as normal as the circumstances allow.[4]

Fermor's portrayal of the asylum was an idealized one based on a short visit. The actual circumstances in Chacachacare were less bright.[5] However, Fermor's perceptions were typical of the images of the modern leprosy asylum among the Western public at the time – images of selfless medical practitioners and religious nurses caring for the medical and social needs of leprosy sufferers.

Half a century later, these images were decisively challenged in the historiography of leprosy. Late twentieth-century historians of medicine regarded colonial and imperial medicine (including leprosy care) as an instrument of social control and a strategy for winning support for colonial rule.[6] To Rod Edmond, the construction of modern asylums should be understood as 'drawing lines, establishing boundaries, constructing enclosures … to contain the dangers of social health and ordering that imperial expansion brought with it'.[7] According to Edmond, leprosy asylums were an example of 'camp-thinking' similar to the British concentration camps in the Boer War and the Native American and Aboriginal reservations in the United States and Australia.[8] Influenced by Foucauldian notions, historians turned from viewing leprosy asylums as institutes of humanitarianism and care to places of repression and isolation. For example, Rita Smith Kipps' investigation of the Lau si Momo asylum on Dutch Sumatra used Ervin Goffman's concept of 'total institution', 'a natural experiment on what can be done for the self'. She found the asylum to be a place where the inmates were colonized and subjected to a European sense of time, space, hygiene, and morality.[9] Australian historian Warwick Anderson, writing about Culion, an asylum in the Philippines, also minimized the role of medical care, treatment, and humanitarian motives. He found that Culion was foremost a colonial reformatory where marginalized people, who were regarded as savages, were transformed in body and mind and changed into citizens, and thus became similar to the bourgeois white males who were considered the end product of civilization. Culion combined confinement with discipline, thus creating the colonial subject who was essential for the exercise of colonial power.[10]

These revisionist historians have situated asylums within the horizon of biopolitics, where biological and medical interventions are an instance of the discourse and execution of power and control in modern society.[11] Whereas Foucault had seen medieval leprosy colonies as an example of sovereign power, exile, and the enclosure of a marginalized group, these historians have suggested that the modern leprosy asylums can be seen as an example of disciplinary power. The cordon sanitaires put around the sufferers were intended not only to isolate them, but also to remake and discipline them. These cordon sanitaires had a distinct racist content.[12] In this revisionist historiography of leprosy, the work of Christian missionaries and their evangelization went hand in hand with the medical drive for social control. The leprosy asylum gave the missionaries the opportunity to further their activities and a place where a new Christian identity could be forged for the isolated leprosy sufferers. From this perspective, it is apparent that there was also friction among the colonizers since the rise of outpatient treatment threatened the missionaries' position, and the shift in location meant that the missionaries lost substantial control over the patients.[13]

Although the changes in the image of leprosy asylums have been profound, approaches of humanitarian improvement and self-sacrifice, and ideas of total institution discipline and control are not entirely irreconcilable. To the doctors, nurses, and missionaries in the asylums, discipline and control were necessary for the patients' best interests. Revisionist historians have primarily questioned this 'best interest of the patient' view, emphasizing instead the view of colonial rule and power strengthening. However, the historians' challenge of traditional images has not gone unanswered as scholarship has focused on asylum patients' agency and the friction, unrest and even resistance towards discipline from above.[14] For example, Jo Robertson has advocated a more nuanced approach to the historical study of the asylums 'against a historical practice that situates the leprosy asylum exclusively within prison-like institutions.'[15] She has shown variations among asylums and the complexities and contingencies in their development in colonial settings. Jane Buckingham has argued that leprosy sufferers' collective isolation on the island Mokagai in Fiji was an enabling experience, thus creating a collective identity against the 'norm' of what was healthy.[16] Leprosy asylums were complex microcosms in which the coercive powers of the colonial state, colonial medicine, missionary societies, and patients from diverse cultural and racial backgrounds came together.

Groot-Chatillon

In Suriname, the cultural and racial diversity was most obvious in the government's Groot-Chatillon, which was administered by the public health service. The director was also the physician in charge. He had to keep control over a patient population of diverse ethnic and religious backgrounds. This was difficult at times because of the sufferers' discontent over religious interference, restrictions on sexual relations, poor food and other aspects of everyday life.

Groot-Chatillon was located on a former sugar plantation on a bend of what was then called the Upper Suriname River in close proximity to the Protestant asylum, Bethesda. The location had several advantages since the breezes from the river made the atmosphere airy and the fresh air and the light were considered beneficial for patients. The terrain of the two asylums was almost completely surrounded by the river and so they were conveniently isolated.[17] Groot-Chatillon provided a place for the compulsorily segregated sufferers, whose families were too poor to finance their treatment and isolation at home. This group included British Indians and Javanese who did not wish to go to or were not welcome in the Christian asylums because they refused to convert to Christianity. Contrary to asylums in India, where the tendency was to admit sufferers of all religious persuasions in the hope of converting them afterwards, the Catholic Majella and the Protestant Bethesda asylums were closed to all but co-religionists in a typical example of Dutch 'pillarization', as discussed below.[18]

During the 'Great Confinement' period of the nineteenth century, almost all of the sufferers in the Batavia asylum had been Afro-Surinamese slaves. Although in the twentieth century the great majority of detected sufferers were still Afro-Surinamese, there was an increased presence of sufferers from new migrant groups, especially the British Indians and to a lesser extent the Javanese who were sent to Groot-Chatillon.[19] By 1902, Groot-Chatillon's population had grown to 149 of whom sixty-four (43 per cent) were British Indians. When including Chinese and Barbadian sufferers, and one European sufferer, almost half of Groot-Chatillon's population at that time was not Afro-Surinamese.[20] That said, the relative composition of the population was in flux, since the colonial authorities had an alternative for segregation for leprosy sufferers from the new migrant groups – repatriation. This could be a convenient way to deal with unproductive labour migrants. In 1905, thirty British Indians were repatriated to India 'at their own request'.[21] Whether this was really at their own request or not is questionable, since they were part of a group that included forty-four patients from the insane asylum and several beggars without means of subsistence.[22] Remarkably enough, the British authorities did not refuse their return. On 1 March 1911, 603 British Indian migrants were shipped for repatriation to India, among whom were thirty-three leprosy sufferers.[23] Javanese sufferers also faced repatriation. In the 1920s, their presence in Groot-Chatillon increased.[24] However, their number dwindled again in the 1930s after Lampe repatriated thirty Javanese sufferers to the East Indies. In 1924, there were twenty-nine British Indians and twenty-five Javanese in Groot-Chatillon in a total of 151 patients.[25] In 1933, there were sixty British Indians and nine Javanese among 221 patients.[26] Almost all the other patients were Afro-Surinamese. This relative distribution continued at least until the end of the Second World War.[27] Groot-Chatillon's ethnic composition meant that it had a sizeable minority of Hindus and Muslims present unlike Majella and Bethesda. For example, on 1 January 1905 there were 141 patients, including fifty-seven Catholics, thirty-one Protestants, and twenty-eight Hindus.[28] Twenty years later, there were seventy-nine Catholics, forty-eight Protestants and twenty-seven Hindus and Muslims.[29]

Since the Christian asylums had insufficient financial means to care for all Christian sufferers, they were present in the government asylum as well. While the colonial state financed the cost of every patient in Groot-Chatillon by paying 300 guilders per year, the Christian asylums of Majella and Bethesda only received 100 guilders for each patient from the state. By 1907, the real cost of patient care in the Christian asylums had risen to 400–500 guilders per patient per year.[30] This imposed limits on the number of patients who could be admitted to the Christian asylums. Therefore, the Protestants from the nearby Bethesda asylum organized religious care and services for the Christian sufferers in Groot-Chatillon, while the Catholics had a priest residing part-time in the government asylum and servicing the surrounding districts.[31] As Groot-Chatillon was meant to be Batavia's successor this included continuing the role of religion in the sufferers' everyday lives. The St Rochus church located in Batavia was moved to Groot-Chatillon, including relocating the saint's statue.[32] At times, religious proselytization was a source of discontent. In 1937, the Public Health Service warned the Catholic Bishop that he was not allowed to spread religious propaganda in Groot-Chatillon. The service had received complaints from British Indian patients that the Catholic priest had tried to baptise Hindu patients on their deathbeds.[33] The next year, the Catholic priest complained that the asylum staff had disturbed his services by organizing musical soirees, film exhibitions, and other activities. He pressed charges with the police, but the director of Groot-Chatillon denied the allegations.[34]

In addition to ethnicity and religion, gender and age contributed to discontent. Most patients were adult males who were more likely to show open resistance to authority than the women or children. During the pre-First World War peak in patient numbers in 1902, 105 of the 149 patients were adult males, thirty-nine were adult females and only five were children. This male dominance in the gender distribution did not change. In 1930, of 158 patients 125 were adult males, thirty were adult females, and three were children.[35] There were some problems caused by the regulation of sexual relations. Healthy spouses were not admitted to Groot-Chatillon, but marriage within the asylum was allowed in contrast to the Christian asylums.[36] There were married couples in the asylum living in small houses. Unmarried patients lived in block dwellings with seven or more to an apartment. The children from the marriages were removed to Paramaribo at the age of two if they tested negative for the bacillus. Extramarital sexual relations were not allowed, but did occur, resulting in the birth of children out of wedlock. In 1922, of the four babies born in Groot-Chatillon and sent to Paramaribo, two were illegitimate. One female patient born in Groot-Chatillon was the daughter of a female patient and a healthy male kitchen servant.[37]

Groot-Chatillon was quite unruly after the first patients were brought over from Batavia. Many of them deplored the loss of freedom of movement compared to what they had enjoyed in the old asylum and their accommodation was rather overcrowded in the beginning. By 1899, general disturbances no longer occurred, but individual patients still expressed their discontent and were punished for their behaviour.[38] To maintain order, a few police officers were stationed on the grounds of the asylum to assist the physician-director, the deputy director (and sometimes head nurse), and the male and female nurses.[39] Unrest at Groot-Chatillon became a topic of discussion in the Dutch Parliament. It was said that the 'garrison life' and discipline in Groot-Chatillon was the major cause of unrest, and that British Indians in particular wished to leave and return to their own country.[40] In 1931, De Vries-Bruins asked in Parliament why the patients in Groot-Chatillon received mouldy bread brought all the way from Paramaribo.[41] A major source of discontent for the Afro-Surinamese patients was the food, which had also been the case in Batavia. British Indians complained less; on festive occasions, such as Santa Claus parties, they were sometimes treated with goat meat.[42] Patients transferred from Majella and Bethesda did not always appreciate their new surroundings and were reported to become excited and even violent. In response, in 1925, nine new isolation chambers were built for unruly patients in addition to the three existing ones.[43] Lampe sometimes used a temporary transfer from other asylums to Groot-Chatillon as a punishment.

A final manifestation of unrest was marronage, the illegal absence from the asylum. Despite Groot-Chatillon's isolated location, more than a few patients escaped. Escapes were sometimes recorded as one a year, but in 1902 there were five and the numbers in subsequent years were eleven in 1908, five in 1910, seven in 1912, twelve in 1913, three in 1914, five in 1917, seven in 1918, four in 1919, and a peak of fifteen escapes in 1920. In 1924, there were another three escapes. The reports do not mention whether these were mass escapes or individual one-by-one exits.[44] Most, but not all, runaways returned of their own accord.[45] In 1927, the medical inspector Cool complained that he had found one escapee selling fruit in the Paramaribo market without being bothered by the police.[46]

Unrest might well have been stimulated by inaction. What were the patients to do the whole day? One solution was to give them work, so in

1916, patients were encouraged to grow bananas and other fruits, and some received permission to breed pigs. This approach had the added advantage of improving the food situation in the asylum, which was welcome since the First World War had disrupted international trade routes and had caused food shortages in Suriname. The patients were allowed to sell the food to the asylum and support their families outside with the money they earned.[47] This might have improved the sufferers' everyday lives. To further improve the situation in Groot-Chatillon, a cinema was installed in 1925. Twice a month the patients could watch a film reel sent from Paramaribo.[48] The next year the patients received a gramophone. They also used their own musical instruments, had a choir, and staged theatre plays.[49] However, although the asylum's facilities were improved, this did not eliminate discontent. After the Second World War, the asylum staff's authority was contested in regard to medical treatment, as detailed later in this chapter. In Groot-Chatillon, the regime of compulsory segregation continued to be contested and permeable.

Majella

Majella was a showpiece of Catholic missionary propaganda, but its segregation regime was as permeable as that of Groot-Chatillon. To the Catholics, the contrast between their new Majella asylum and Batavia was total.[50] Batavia could only be reached by a long journey over ocean and river to the jungle and had become decrepit. Majella was situated on the outskirts of Paramaribo and looked quite distinct. The terrain was dominated by two white caulked rows of clean buildings opposite each other: one for male patients and one for female patients. Moreover, this patient population was relatively homogeneous compared to Batavia and Groot-Chatillon, since patients had to be Catholic or convert to Catholicism. Gender and age distribution was more consistent than in Groot-Chatillon. In 1910, of 128 patients only a minority (forty-seven) were adult males, and thirty were adult females. The number of children was much larger: fifty-one, including fifteen girls. In 1930, the number of adult females was even greater than adult males: fifty-five compared to fifty-two, with eighty-seven children (including thirty-two girls).[51] The number of nurses caring for these patients rose to twenty-six nuns of the Congregation of the Sisters of Tilburg by 1935. By then, most nurses (seventeen) had been in the asylum for twenty years or more.

Under the first director, Father Felix Lemmens, a former soldier who had become a priest in 1886, a strict time schedule of religious activities was devised that regulated the inmates' lives starting with morning prayer in church at 6.30 a.m., and continuing with mass at 7.00 a.m., catechisation lessons during the day and evening prayer at 7.00 p.m. This regime was aimed at increasing the patients' devotion focused on the suffering of Christ and Mary, in her role as Our Lady of Sorrows.[52] Catholic propaganda reported favourable conditions in the playing grounds, the garden, and the roomy and airy building that housed the kitchen, laundrette and clothing-repair facilities. Every patient older then fourteen had his or her own apartment. Between 1927 and 1933, new apartment blocks were built for men and women, as well as a larger playing ground for the children, and a new house for the nurses.[53] If capable, the women assisted the nurses in their everyday labour, while capable males did woodwork or practised agriculture. In 1937, the Bishop thought that the patients could do more work, such as gardening, clothing and shoe repair, construction and woodwork, which would be 'pleasant and useful' labour that could financially benefit the asylum.[54] This was a response to budget cuts in the government subsidies for patients because of the economic crisis, but it was presented as an amelioration of conditions of life in the asylum.[55] In addition to work, opportunities for recreation included choirs, theatre performances, weekly concerts, and sports such as football, cricket, and korfbal, a Dutch version of basketball. Religious activities gave succour as well as discipline to the everyday lives of the patients.

Not everybody in Paramaribo subscribed to the almost idyllic Catholic image of life in Majella. One cause of resentment against the Catholic asylum was its location on the outskirts of Paramaribo. The inhabitants of Paramaribo were anxious about the possibility that objects used by the patients in Majella could be contaminated with leprosy and might find their way to the patients' families and friends in town thus spreading the disease. In 1902, a member of the Dutch Parliament asked for the asylum to be relocated. He was particularly concerned that the patients washed laundry for and sold cakes to people outside the asylum.[56] In 1914, the Colonial Estates had to declare once more that Majella's location was not a risk to the health situation in town.[57] New complaints from inhabitants in Paramaribo were registered the next year.[58] These complaints did not hinder the construction of a concrete bridge over the Sommelsdijckse Creek that connected Majella with the town in 1919.[59] To counter anxieties, in 1926, the Bishop admonished the sisters of Majella that objects should not leave the asylum.[60] In 1928, the Colonial Estates declared that patients who went to Majella because of its location close to town should not be forced to go to a more isolated asylum. They believed that the only danger of contagion was close physical contact with patients. Rats and cats could possibly transmit the contagion, but not insects.[61]

Criticism continued since patient segregation in Majella was as permeable as it was in Groot-Chatillon. In September 1931, Lampe complained to the Bishop that healthy children and young adults freely moved in and out of Majella for patient visits. He asked for a stricter police control of asylum visits and regular check-ups of the situation every one or two months.[62] In 1932, Lampe requested that children younger than seven years not be allowed to visit the asylum, while those between seven and thirteen years should only visit once a month on the regular visiting day, and patients with severe signs of leprosy should not be allowed visitors.[63] In 1934, the attorney general complained that patients went into town on their own initiative. For the attorney general, the separation between Majella and the town was inadequate.[64] On the Queen's birthday, patients were allowed to go into town and participate in the festivities.[65] In 1935, the Public Health Service sent no less than three requests to the Bishop to complete building a fence between Majella and a football field used by neighbourhood residents. According to medical inspector Cool, young male patients in particular went out unchecked at night through the football field into town. However, the Bishop declared that he did not have the financial means to complete the fence.[66] Cool also complained that even female patients went into town.[67] Objects belonging to patients found their way into town without being decontaminated.[68] The director of Majella did not know how to handle the situation and asked Schuitemaker for advice on how to manage young patients who regularly went into town without permission. Schuitemaker reported this to the police and in response the police questioned the Bishop about the situation. The Bishop preferred to keep the situation a secret and became angry with the director.[69] Cool also asked for the construction of an isolation chamber for 'unruly or difficult elements' among the patients.[70]

Complaints about the asylum from the Public Health Service continued throughout the 1930s and 1940s, such as visitors bringing food to patients at irregular hours, and objects (even a radio transmitter) disappearing from the asylum without prior decontamination.[71] The consequence of Majella's permeability was that contrary to the opinions expressed by the Bishop and the Colonial Estates, doctors worried about the possible transmission of contagion from the asylum into town. Leprologist S. J. Bueno de Mesquita suggested in the 1940s that the incidence of leprosy in a circle of two kilometres around Majella was twice the incidence of the disease in other parts of Paramaribo.[72] Furthermore, there was criticism of the health situation in Majella. The leprologist Keil, who was in charge of the asylum's medical supervision in the first half of the 1930s, complained about non-existent filariasis prevention, inadequate facilities to conduct medical examinations, violated diet regulations, and nurses not being persuasive enough with the patients about submitting to treatment.[73]

Although the patient population of Majella was more homogeneous than in Groot-Chatillon, and patients were subjected to a regime of strict Catholic discipline, this regime was permeable and there was relative freedom of movement between the asylum and Paramaribo. Because of this close proximity, contestation of the segregation regime took the character of a petit marronage, if only for one night or a few hours, rather than a grand marronage or permanent flight as in the more isolated Groot-Chatillon.[74]

Bethesda and New-Bethesda

Just as Majella had been, Bethesda was used as a showpiece in missionary propaganda; however, this time by the Protestants. Bethesda was situated next to Groot-Chatillon, separated only by a canal.[75] Protestant propaganda attacked the alleged Catholic's use of Majella's location close to Paramaribo to lure patients to the asylum, even if this meant conversion to Catholicism. Protestant patients who preferred to go to the jungle and stay with their religious brethren in Bethesda were praised.[76]

Protestant propaganda was needed to finance Bethesda with contributions from co-religionists in the Netherlands and later the United States. Financial contributions from Suriname were limited; for instance, in 1909 only approximately 2,000 guilders were donated to address almost 45,000 in expenditures.[77] Most of the concerns of the Dutch headquarters of the Moravian mission in Zeist were about collecting funds for and financing of Bethesda.[78] In 1902, Zaalberg founded a Central Committee in the Netherlands to organize financial support for the Protestant Association that ran the leprosy asylum. The first quarter of the year after the committee was formed, more than 60,000 guilders were collected for the asylum.[79] By 1925, the Committee had almost 1,500 regular financial contributors.[80] The support of the Dutch royal family was of significant importance in gathering support for Bethesda in the Netherlands. After the Queen Mother Emma's death in 1934, her daughter, Princess Juliana, the heir to the throne, became the patroness of the Protestant leprosy asylum.[81] Outside the Netherlands, another important source of revenue was the Bethesda Leper Home Society in Buffalo, New York in the United States. In addition to government subsidies, the Americans became the chief financers of the asylum after the First World War.[82] The German Moravians were another main source of support. The nurses, nine by the time Bethesda was relocated in 1933, were Protestant deaconesses and in the beginning most were sent by the Moravian Church in Niesky, Germany.[83] The asylum directors and preachers or 'teachers' were also Moravians and until the 1930s, the director's salary was paid for by funds from Germany.[84] German involvement in Bethesda would only end after the Second World War. In 1947, the last three German deaconesses left Suriname and were replaced by Dutch ones.[85]

At first, the asylum was quite small, with a Protestant church, a house for the Moravian director and preacher or 'teacher' and his family, and two little houses for patients.[86] In 1908, more buildings were constructed: a large house for the deaconesses, a big kitchen, and new apartments for the patients. There were separate houses for male and female patients. In 1924, a building for children was constructed.[87] The gender distribution in Bethesda was similar to Majella. In 1910, there were fifteen adult males and seventeen adult females among the forty-two patients. In the 1920s, the gender distribution leaned somewhat towards male patients, but did not become as unevenly distributed as in

Groot-Chatillon. In 1930 there were thirty-eight adult males and eighteen adult females among the eighty-seven patients.[88]

By the end of the 1920s, the Bethesda asylum had become rather overcrowded. Until then, the patients had had more living space than in the other asylums. However, Bethesda's location far from town was much less popular than Majella. In 1933, Bethesda moved to a new location. In the new leprosy politics after 1929, with mandated outpatient treatment and clinics, it was no longer of paramount importance that an asylum be located far from the civilized world. When the asylum became overcrowded, and the river further eroded banks of Bethesda thus threatening the nurses and director's houses, leprologist Keil supported the request to move the asylum to a new location in Paramaribo. At first the Protestant Association for Leprosy in the Netherlands that operated Bethesda disliked the idea of relocation and was of the opinion that Dutch expertise in managing floods could solve the situation. However, Flu and other medical practitioners believed that treatment close to town was preferable to segregation in the jungle and they were able to sway the Association. Moving the asylum had the added advantage that the problem of illegal contact between patients in Bethesda and Groot-Chatillon would also be solved.[89] In 1932, the government allocated a loan without interest of 70,000 guilders to Bethesda to build a new asylum to be called New-Bethesda (Nieuw-Bethesda).[90] On 3 November 1933, Governor Kielstra officially opened the new asylum.[91] It was in a much more favourable location than the old asylum, since it was on grounds four miles outside of Paramaribo with a view of the city. There were twenty-seven white painted buildings with red-tiled roofs on the grounds segregated from the neighbourhood on a large deforested terrain along the road to Domburg.[92] New-Bethesda had significantly more space than Bethesda had and could care for three times as many patients (180 instead of approximately sixty).[93] As in Majella, a cinema was installed.[94] However, not all was favourable; for example, the grounds of New-Bethesda were plagued by mosquitoes.[95] At first, malaria was rampant, until the introduction of new medication received for free from I. G. Farben that slowly stabilized the situation.[96] New-Bethesda's close proximity to Paramaribo also meant that people visited non-existent patients out of curiosity to see the asylum or to sell goods.[97]

The everyday regime in Bethesda and New-Bethesda was as strict as in Majella with morning and evening prayers and daytime activities in between. Patients who wished to do so could work in gardening, pig or chicken breeding, shoe making or assisting nurses in their everyday chores. After the asylum's relocation to New-Bethesda, it took time for new patients to begin working in gardening and other activities. According to the nurses, the patients had to be educated that labour was a blessing and not a curse, and to show their gratitude as had been the situation in the old asylum.[98]

Another problem for the staff was language. Since the German nurses at first did not understand Dutch or Sranan Tongo, there was a language barrier with the patients. Finally, in 1930, the German nurses sent by the Moravians took a course in tropical hygiene in Amsterdam and learned the Dutch language before they went to Suriname.[99] However, a cultural barrier remained between nurses and patients. The missionaries disliked the obvious solution of appointing Surinamese-born nurses, as suggested by Cool. Echoing the racist prejudices of colonial society, they thought that these nurses might have had the right formal education, but were lacking the correct Christian spirit. The missionaries were relieved when four new sisters came from Europe to New-Bethesda in 1936. In their view, Surinamese women should only be appointed as assistant nurses.[100] When the number of patients rose to almost full capacity (172) in 1938, four European sisters were assisted by two Surinamese.[101]

Food was a source of unrest. As in other asylums, food became expensive. Since most of the nurses and much of the financial support came from Germany, the First World War disrupted the flow of resources. For years after the war, Bethesda had financial difficulties.[102] The food shortages stimulated Bethesda to use the relatively healthy patients more for work in agriculture and animal husbandry.[103] Another problem was the insufficiency of the water supply, making the asylum completely dependent on rain and leading to shortages in periods of intense drought.[104] There were also shortages of bandages that then had to be collected in the Netherlands.[105]

From the very beginning, Bethesda emphasized the teaching of Christian virtues to the patients: thrift, gratitude, and contentment, brotherly and sisterly love, and communion with each other and with

God.[106] Not all patients appeared to have been Protestants. By 1935, the asylum had six Hindu patients and three Catholics.[107] It is possible that these patients came there on the condition of conversion to Protestantism. In 1954, when the ten-year-old Hindu boy Eddy Jharap was diagnosed with leprosy in the outpatient clinic in Paramaribo, he was sent to Bethesda and had to participate in Christian activities: saying prayers, going to church, and attending sermons on Jesus' special devotion to leprosy sufferers.[108] Even if a patient was not a Protestant, in Bethesda all patients were subjected to strict Protestant religious discipline.

Not all patients obeyed the religious commandments. Bethesda was not keen on admitting patients from the other asylums out of fear that they would not adhere to the strict Protestant regime and discipline.[109] Staff were in fear of a situation that had occurred in the St Louis asylum just over the border in French Guiana. St Louis was rumoured to be an asylum that was filled with 'money, gold, alcohol, cards, dice, opium'.[110] According to the staff, in Bethesda most patients adjusted to the regime and started to look upon the asylum as their home. For some, whose 'memories of a dissolute life were not seldom their only treasure', this took a bit longer than for others.[111] To the staff, the Sunday school was essential in inculcating the right virtues in the patients. Patients that still remained unruly faced being transferred to Groot-Chatillon.[112] In the first quarter-century of Bethesda's existence, no less than thirty-two patients were removed to the government asylum, probably for disciplinary reasons.[113]

Patients contested the regime in Bethesda in various ways. Bethesda's separation from Groot-Chatillon by a canal was far from complete and it appeared that there was regular (although prohibited) comings and goings between patients from Groot-Chatillon and Bethesda. According to the Bethesda staff this led to 'fornication and theft'.[114] Although marriages and sexual relationships were prohibited, just as in Majella, sexual intercourse between patients did occur. One female patient who had been transferred to Groot-Chatillon was discovered in a man's apartment at night. Other patients who had been transferred included a couple who moved to the government asylum where they were allowed to marry.[115] In terms of imposing religious discipline on the patients and the contestation of this discipline in everyday life, there was little difference between the Protestant and the Catholic leprosy asylums.

Medical treatments

From the 1900s onwards, colonial medical practitioners experimented with methods of treatment for leprosy sufferers inside and outside of the asylums. Although the efficacy of these methods was questionable until after the Second World War, they were essential for colonial medicine's perception of what a modern asylum should be – a place not only for care or discipline but also potentially for cure. Patients were encouraged to participate in these treatments, but could not be forced. A closer examination of the medical treatments suggests that in this area, sufferers could exercise some degree of agency and control over their everyday lives, and acquire a value of self in making their own life decisions.

As a government asylum run by the Public Health Service, Groot-Chatillon became the centre of medical experiments. In 1903, experiments were already being undertaken to treat three patients with the tua-tuaheester or bellyache bush (Jatropha gossypifolia). These experiments continued for two years until the plant's supply was depleted. The results were indifferent. There were no cures and only in two of the three patients was there a partial improvement in their general condition after one year of treatment. In 1905, other experiments were started with a medicine developed by a Parisian physician, Dr Brinson, using mangrove bark, and in 1906, with intravenous injections of chaulmoogra oil in combination with strychnine. Chaulmoogra was injected because when it was administered orally it caused patients to have severe nausea. Treatment with this drug succeeded in arresting and even reducing the symptoms of six patients after one year of treatment. However, in 1911, the results were described as 'discouraging' since patients could not bear the treatment. Aiouni therapy turned out to be unsuccessful as well. Here, and with other medication, patients' enthusiasm to participate in the experiments was low, primarily because it meant going on an adjusted and impoverished diet. For instance, in 1908, there were no patients who volunteered for experiments with a new German pharmaceutical medicine called Nastin-B (this drug was injected and consisted of nastin, a fat extracted from the leprosy bacillus, combined with benzochloride).[116] The director of the Military Hospital in Paramaribo Koch made use of the close proximity of Majella to experiment there with another treatment – injections of mixtures of iodine and olive oil. Father Lemmens, who had become a leprosy patient himself in 1902, was one of the volunteers for these unsuccessful experiments.[117]

By 1915 preference was again given to the administration of chaulmoogra oil in Groot-Chatillon. Until the rise of sulfone therapy after the Second World War, this medication was the height of international medical fashion in the treatment of leprosy. British doctors, such as Ernest Muir, claimed credit for the development of chaulmoogra therapy, but Dutch doctors in Suriname insisted that they had developed their treatment method independently. Schuitemaker imported various kinds of chaulmoogra oil from the United States.[118] Oral administration of chaulmoogra oil in combination with gelatine helped improve patient acceptance by diminishing the problem of nausea.[119] In 1916, thirty patients received this treatment. A few experienced gastric disturbances.[120] Therefore, the doctors returned to intravenous injections of chaulmoogra with olive oil and camphor for these patients. On the whole the doctors were not completely dissatisfied with the results, but not very enthusiastic about them either. In December 1917, chaulmoogra oil treatments were temporarily halted because of an influenza epidemic.[121] In 1919, the treatment was offered again in Majella.[122] In 1921, the Colonial Report described chaulmoogra oil treatments in Groot-Chatillon and Bethesda.[123] Despite the initial enthusiasm (sixty-three patients volunteered for treatment), the results were discouraging and within a few months half of the volunteers had withdrawn from treatment.[124] In 1924, the method was again changed and chaulmoogra was now intravenously injected in the form of antileprol (a preparation marketed by I. G. Farben), which was considered less painful when combined with cycloform.[125] However, patients continued to prefer to have their bodies anointed with the chaulmoogra oil or take an oral administration.[126] Treatments with antileprol were also started in Majella (sixty-two patients) and Bethesda (twenty-seven patients).[127] The results of chaulmoogra therapy made Lampe and other doctors in Suriname optimistic about the possibilities of the method and the execution of a modern leprosy politics that promised treatment and not only segregation.[128] In the 1928 annual report of the Public Health Service, Lampe declared that in that year twelve patients had been cured, fifty-nine showed much improvement, twenty-one were repeatedly bacteriologically negative after always having tested positive, and 131 patients showed no change, while one had died. The treatment consisted of intramuscular injections of alepol (a preparation of chaulmoogra marketed by Burroughs Wellcome) twice a week, daily oral administration of chaulmoogra oil, 'as much as one can bear', or daily ointments of the body with the oil, or warm baths with chaulmoogra soap. Alepol was used in the outpatient clinic as well as in the asylums.[129] Many patients first had to be treated for other diseases as well: yaws, syphilis, or malaria. In 1929, twenty-four patients were declared cured.[130]

A letter from Cool in 1933 casted doubt about the strictness and regularity of the treatment, at least in Majella. He asked for more information from the Bishop about rumours that when patients asked for antileprol with cycloform injections they sometimes received another medication. There were complaints about burning spots on the patients' bodies with the use of trichlorarsenic. The physician responsible, Keil, was not given to talk to the patients and improve their mood. He did not examine their constitution and just treated them instead. Patients received a diet without salt, meat, or fish.[131] Cool's negative opinion about treatments in Majella might have been justified. In 1936, R. Vervoort, a man who claimed to have found an expensive medicine against leprosy, was given permission by the Bishop to experiment in Majella if he did it for free.[132]

Patients in Majella and Bethesda were more disciplined than in Groot-Chatillon. Volunteers for medical treatments in the two Christian asylums rarely failed to turn up for treatment sessions.[133] Elsewhere, the patients who were less pressured by religious discipline and could not legally be forced to participate in treatments could and did make their own treatment decisions. In 1909, for a short time, suspects of leprosy were admitted to the Military Hospital and treated with chaulmoogra. Two patients requested and received their own treatments, although these also failed. One Afro-Surinamese suspect preferred some unspecified herbs from folk medicine, and a German suspect preferred treatment with Nastin-B.[134] In 1924, of sixty-six patients undergoing treatment in Groot-Chatillon, forty-three did not turn up for their injections at least once a month. In 1926, nineteen patients refused all further treatment. This confirmed the physician's prejudices. He was of the opinion that these people were completely indifferent to improvements in their situation.[135] According to the Public Health Service, Afro-Surinamese patients,

Do not want to be treated, because of the fear, that after being allowed to leave the infirmary, they will not have a proper home. Others, who have mutilated hands and feet, also fear the struggle for life in the city, even if they do have a home at present where they can go to.[136]

Up until the end of the Second World War, chaulmoogra therapy continued. In 1946, leprologist Bueno de Mesquita started switching to the new and much more effective sulfone therapy. He began in New-Bethesda with daily intravenous injections of promin (sodium glucosulfone). At first, patients had to pay for the cost of the drug. Eventually, a private fund was organized and contributions from Surinamese who had moved to the Dutch Antilles, and the Netherlands were collected. Some support was received from postage stamp sales by the colonial government. This fundraising made it possible to treat more patients with the drug. Unlike chaulmoogra, promin caused no adverse side effects in patients, apart from a 'leprosy reaction', in which symptoms of the disease temporarily increased. Although the bacteriological effect was slow, most patients showed therapeutic progress. Spots disappeared, nose bleeds ceased, and hair and nails grew back. Because of the high costs of promin, the use of diasone (another sulfone drug) was introduced in 1947. A total of 30,000 tablets were received from the American Mission to Lepers. Diasone was orally administered. There were more side effects than with promin, such as headaches and depression, and patients had a quite severe leprosy reaction although the reactions diminished over time. After one to two years of therapy, sometimes first with promin and then diasone, the subjective and objective results of the use of sulfones were quite promising. Unlike chaulmoogra therapy, there were no patients whose condition deteriorated, thus chaulmoogra therapy was ultimately discontinued in 1948.[137]

In Groot-Chatillon, patients continued making their own decisions about treatment. When Bueno de Mesquita administered sulfone in Bethesda, the patients in Groot-Chatillon felt left out. They did not receive the drug possibly because it was too expensive or the asylum was located too far away for the government leprologist from Paramaribo to oversee the treatments on a daily basis. In 1948, patients in Groot-Chatillon read about promin in *Time* magazine and went to the physician-director of the asylum. One of the patients remembered how the director obtained a limited supply of the drug and successfully treated one patient. Other patients also received the drug when their families could afford to buy it for them. However, this led to a revolt among the poorer patients who did not receive the drug. Eight poor patients who were still able to walk escaped the asylum by boat and walked thirty-five kilometres to Paramaribo in order to complain to the press and the Colonial Estates, before turning themselves in. The patients were sent to jail in Groot-Chatillon, but were freed by their fellow-patients. Finally, they achieved their goal and promin was administered to all patients in Groot-Chatillon when it became available a few weeks later.[138] Patients showed that medical authority in the government asylum was contested and that they were willing to make their own decisions and act for their own benefit.

Conclusion

Leprosy asylums in Suriname were only to a certain degree rigid disciplinarian institutions, despite the intentions of their founders, the religious missions, and colonial medicine. The segregation regimes were characterized by varying degrees of permeability. It is true that all the asylums had their own programs of discipline and punishment for their patients. The Christian asylums were organized around a strict observance of religious duties devised to inculcate a Christian identity and gratefulness in the sufferers. There was also a Christian presence in the state asylum. Missionaries and colonial doctors worked together, despite frictions about the control and execution of treatments. However, the patients in the asylums succeeded in exerting agency to a greater or lesser degree and contesting the directors, staff, and doctors' authority. The patients had their own infrapolitics of dealing with the disciplinary regimes. This was possible in part because of the many practical problems of controlling everyday life in the asylums. In Groot-Chatillon, there was a relatively heterogeneous population of various ethnic and religious backgrounds and a majority of patients were adult males. The patients resented interference in matters of religion and sexual relations and rules were disobeyed. Though isolated far from town, illegal visits to the Bethesda asylum next door and escapes from the asylum in general could not be prevented. While Majella and Bethesda had more homogeneous populations and religious discipline was strictly enforced, contested authority and temporary escapes existed here as well. Marronage was not uncommon, whether *petit* (escaping from Majella for one night to go into town), or *grand* (escaping permanently from Groot-Chatillon). In regard to medical treatment, sufferers could and did exercise some degree of agency and control over their everyday lives. Most patients did not wish to submit to medical treatments often perceived as painful and useless and they were not forced to. Asylum staff and nurses were less than enthusiastic about encouraging medical treatment, and were sometimes unable to cope with disobedient behaviour.

Patients' lives while in the asylums were complex, since various agendas had to be negotiated and multiple groups had their own instruments for negotiating everyday life. The leprosy sufferers might have become compulsorily segregated, taken from their families and friends and isolated in a new environment, but they had not lost all agency or the capacity to act.

Notes

1 R. Smith Kipp, 'The evangelical uses of leprosy', *Social Science and Medicine* 39 (1994), pp. 165–78; W. Anderson, 'Leprosy and citizenship', *Positions* 6 (1998), pp. 707–30; S. L. Burns, 'From "leper villages" to leprosaria: Public health, nationalism and the culture of exclusion in Japan', in C. Strange and A. Bashford (eds.), *Isolation: Places and Practices of Exclusion* (London: Routledge, 2003), pp. 104–18; W. Anderson, *Colonial Pathologies: American Tropical Medicine, Race and Hygiene in the Philippines* (Durham: Duke University Press, 2006), pp. 158–79; R. Edmond, *Leprosy and Empire: A Medical and Cultural History* (Cambridge: Cambridge University Press, 2006).

2 E. Silla, *People Are Not the Same: Leprosy and Identity in Twentieth Century Mali* (Heinemann: Portsmouth, 1998); J. Robertson, 'The leprosy asylum in India, 1886–1947', *Journal for the History of Medicine and Allied Sciences* 64 (2009), pp. 474–517; J. Buckingham, 'The inclusivity of exclusion: Isolation and community among leprosy-affected people in the South Pacific', *Health and History* 13 (2011), pp. 65–83; K. A. Ingliss, *Disease and Displacement in Nineteenth-Century Hawai'i* (Honolulu, HI: University of Hawai'i Press, 2013).

3 P. L. Fermor, *The Traveller's Tree: A Journey through the Caribbean Islands* (New York: New York Review Books, 2011 [1950]), pp. 160–1.
4 Fermor, *Traveller's Tree*, p. 161.
5 On Chacachacare: D. McCollin, 'Chacachacare: The island of lepers, 1922–1979', in C. Bonfield, J. Reinarz, and T. Huguet-Termes, *Hospitals and Communities, 1100–1960* (Oxford: Peter Lang, 2013), pp. 263–90: 'the forced confinement, the restrictions of religious life and the insulation from the outside world produced an artificial community that was less than ideal for the patients and increasingly untenable for the staff' (p. 263).
6 For example, see D. Arnold (ed.), *Imperial Medicine and Indigenous Societies* (Manchester: Manchester University Press, 1988).
7 Edmond, *Leprosy and Empire*, p. 187.
8 Edmond, *Leprosy and Empire*, pp. 183–96.
9 Smith Kipp, 'Evangelical uses of leprosy', pp. 166–7.
10 Anderson, 'Leprosy and citizenship'; Anderson, *Colonial Pathologies*, pp. 158–79.
11 Burns, 'From "leper villages" to leprosaria'.
12 A. Bashford, *Imperial Hygiene: A Critical History of Colonialism, Nationalism and Public Health* (Basingstoke: Palgrave Macmillan, 2004), pp. 81–113.
13 Smith Kipp, 'Evangelical uses'. On the role of missionaries, see also M. Vaughan, *Curing their Ills: Colonial Power and African Illness* (Cambridge: Polity Press, 1991); S. Kakar, 'Leprosy in British India, 1860–1940: Colonial politics and missionary medicine', *Medical History* 40 (1996), pp. 215–30; M. Worboys, 'The colonial world as mission and mandate: Leprosy and empire, 1900–1940', *Osiris* 15 (2000), pp. 207–18.
14 Kakar, 'Leprosy in British India'; W. U. Eckart, *Medizin und Kolonialimperialismus: Deutschland 1884–1945* (Paderborn: Schöningh, 1997); S. Au, *Mixed Medicines: Health and Culture in French Colonial Cambodia* (Chicago, IL: The University of Chicago Press, 2011).
15 Robertson, 'Leprosy asylum in India', p. 474.
16 Buckingham, 'Inclusivity of exclusion'.
17 F. P. Schuitemaker, *De lepra en de gouvernementsleproserie in Suriname* (Amsterdam: J.H. de Bussy, 1915), pp. 4, 20–2.
18 For India, see J. Buckingham, *Leprosy in Colonial South India: Medicine and Confinement* (Basingstoke: Palgrave, 2002).
19 Colonial Reports (hereafter CR) 1909, Appendix M.
20 CR.

21 National Archive, Paramaribo, Gouvernementssecretarie Suriname 1829–1954 (hereafter GS), 1228, 'Ingekomen rapport voor het Koloniaal Verslag over 1905'.
22 CR 1906.
23 CR 1912, Appendix L.
24 Archive Bisdom of Paramaribo (hereafter BP), T 148.
25 CR 1925, p. 20.
26 GS 1256.
27 GS 1263.
28 GS 1228, 'Ingekomen rapport voor het Koloniaal Verslag over 1905'.
29 CR 1926.
30 GS 273/18, Governor Idenburg in session Colonial Estates 5 March 1907.
31 BP T178.
32 A. C. Schalken, 'Historische gids bestaande uit chronologische lijst naamlijsten varia registers. 300 jaar R.K.-gemeente in Suriname 1683–1983' (Paramaribo, 1985), p. 55.
33 BP T 209.
34 BP T 218.
35 CR.
36 Schuitemaker, *Lepra*, pp. 31–6.
37 BP T 148.
38 CR 1899, p. 13; Schuitemaker, *Lepra*, p. 22.
39 Schuitemaker, *Lepra*, p. 11.
40 J. M. Plante Fébure, *West-Indië in het parlement 1897–1917, Bijdrage tot Nederland's koloniaal-politieke geschiedenis* ('s-Gravenhage, Martinus Nijhoff, 1918), p. 53.
41 *Handelingen Staten-Generaal*: www.statengeneraaldigitaal.nl/ (hereafter SG): Handelingen Tweede Kamer Staten-Generaal 18 February 1931.
42 GS 1237, 'Ingekomen rapport voor het Koloniaal Verslag over 1912'.
43 CR 1926, p. 25.
44 Figures from CR.
45 GS 1254.
46 P. Cool, 'Regeeringsmaatregelen ter bestrijding der lepra in Suriname en Aruba', *Nederlandsch Tijdschrift voor Geneeskunde* 71 (1927), pp. 2453–6, on p. 2453.
47 GS 1243, Ingekomen rapport voor het Koloniaal Verslag over 1916'; Colonial Report 1918; GS 1246, 'Ingekomen rapport voor het Koloniaal Verslag over 1918'; CR 1919.
48 CR 1926, p. 25.

49 CR 1927, p. 26.
50 A. Verheggen, *God is liefde, Veertig jaren melaatschenverpleging 1895-1935* (Maastricht: Gebr. Van Aelst, 1935); H. B. Dresen, *Gouden jubilé van de R. K. leprozerie Sint Gerarus Majella Stichting* (Paramaribo: Sint Gerardus Majella Stichting, 1945).
51 Figures from CR.
52 Verheggen, *God is liefde*, pp. 40-51. On Lemmens: R. Lampe, 'Ik werd militair' (n.p., 1947); L. C. van Panhuys, 'De opoffering van een R.K. priester in Suriname', *West-Indische Gids* 18 (1937), pp. 201-6; J. Vernooij, 'Een opvallende relatie. De rooms-katholieke kerk en lepra in Suriname', *OSO* 22 (2003), pp. 62-8, on p. 65; G.-J. Hallewas, 'De gezondheidszorg in Suriname' (Unpublished PhD thesis, Groningen University, 1981), p. 43.
53 Schalken, 'Historische gids', pp. 95-102.
54 BP T 191, 201.
55 SG: Handelingen Tweede Kamer der Staten-Generaal 1/5, voorlopig verslag over begroting voor 1934, 16 January 1934.
56 Plante Fébure, *West-Indië in het parlement*, vol. 2, p. 825.
57 BP T 179.
58 BP T175.
59 Schalken, 'Historische gids', p. 81.
60 BP T154.
61 Vernooij, 'Barmhartigheid', p. 20.
62 BP T 175.
63 BP T175.
64 BP T 184.
65 Cool, 'Regeerings-maatregelen', p. 2453.
66 BP T189.
67 BP T 207.
68 BP T 206.
69 BP T199.
70 BP 196, letter Cool to the Bishop, 22 February 1937.
71 BP T 228.
72 H. Menke, interview with Paul Niemel, 'Lepra in Suriname: van segregratie naar integratie', *OSO* 22 (2003), pp. 21-33, on p. 27.
73 BP T 179.
74 On petit and grand marronage: G. Debien, 'Le marronage aux Antilles françaises au XVIIIe siècle', *Caribbean Studies* 6 (1966), pp. 3-43.
75 *Stemmen uit Bethesda* (Amsterdam: De Bussy, 1900-1951) (hereafter SB) 1 (1900), p. 6.

76 SB 18 (1912), p. 35.
77 SB 14 (1909), p. 17.
78 Archive Zeister Zendingsgenootschap (ZZ) Utrechts Historisch Archief, Utrecht, The Netherlands (hereafter ZZ) 48/1, 1233, 1234.
79 *Verslag der herdenking*, pp. 9–13.
80 SB 32 (1926), p. 5.
81 *Vijftig jaren Protestantse Melaatsen verpleging "Bethesda" in Suriname 1899–1949* (Paramaribo: Protestantsche Vereeniging, 1949).
82 *Verslag der herdenking*, pp. 9–13; SB 29 (1923), p. 19.
83 SB 37 (1931), p. 28.
84 Necrology in SB 28 (1922), pp. 27–37; salary director: SB 34 (1928).
85 SB 51 (1948), p. 7.
86 For pictoral images of Bethesda and later New-Bethesda: *Bethesda. Een liefdewerk der protestanten in Suriname* (Paramaribo: Protestantsche Vereeniging ter verpleging van Lepralijders in de Kolonie Suriname, 1902); *Stemmen uit Bethesda (SB)* 1–56 (1900–1951); I. Stern, *Sternen-Saat* (Curitiba, 2010).
87 CR 1925, p. 20.
88 Figures from CR.
89 SB 37 (1931), pp. 5–10, 32.
90 SG: Handelingen Tweede Kamer der Staten-Generaal 25 February 1932; 'Tweede nota van wijzigingen van de begroting', 1 November 1932; SN: Keil in *De West* of 6 November 1933, BP T 182; R. D. G. P. Simons, 'De maatschappelijke beteekenis der Surinaamsche ziekten', *West-Indische Gids* 14 (1933), pp. 429–39, on p. 435.
91 SB 40 (1934), p. 10.
92 *Brieven uit Bethesda (BB)*, 1–6 (1934–1937); SB 38 (1932). Pictoral images of New-Bethesda: M. Weigel, *Brieven uit Bethesda* (Paramaribo: Nieuw-Bethesda, 1934); *Veertig jaren Protestantsche Melaatschen verpleging "Bethesda" in Suriname* (Paramaribo: Protestantsche Vereeniging, 1939).
93 B. J. C. Reijnders, 'Het werk van de Evangelische Broedergemeente in Suriname', *West-Indische Gids* 28 (1947), pp. 300–11, on pp. 309–11.
94 SB 38 (1932), pp. 27–8.
95 SB 40 (1934), p. 56.
96 ZZ 48/1, map 111; SB 3 (1935).
97 SB 38 (1932), pp. 26–7.
98 SB 38 (1932), pp. 24–5.

Asylums and treatments, 1900–1950

99   SB 37 (1931), p. 11.
100 ZZ 48/1, 1236, letters Siegfried Beck, Paramaribo, 4 January 1936; 11 April 1936.
101 SB 45 (1939), p. 7.
102 SB 24 (1918), p. 20; J. Postma, 'De leproserie Bethesda tussen 1897 en 1928', *OSO* 22 (2003), pp. 69–81.
103 SB 25 (1919), p. 19.
104 SB 32 (1926), p. 120; 40 (1934), p. 57.
105 SB 32 (1926), pp. 63, 69.
106 'Onder de Melaatschen te Groot-Chatillon', *Berichten uit de Heidenwereld* 1900, p. 4.
107 SB 42 (1936), p. 8.
108 E. Jharap, *Vertrouwen in eigen kunnen* (The Hague: Amrit, 2007), pp. 45–57.
109 SB 23 (1917), p. 16.
110 'Jaarverslag van de Zending in Suriname', *Berichten uit de Heidenwereld* 1910, p. 139.
111 SB 27 (1921), p. 26.
112 Jaarverslag, *Berichten uit de Heidenwereld* 1910, p. 138; B. la Trobe, 'Bethesda en zijn zendingsarbeid onder de Lepra-lijders in Suriname', *Berichten uit de Heidenwereld* 1918, p. 4.
113 *Verslag der herdenking*, p. 30.
114 SB 38 (1932), p. 32.
115 Postma, 'Leprozerie Bethesda', p. 72.
116 GS 1228, 'Ingekomen rapport voor het Koloniaal Verslag over 1905'; CR 1904; CR 1906; CR 1907; CR 1908; CR 1909; 'Ingekomen rapport voor het Koloniaal Verslag over 1912', GS 1237; CR 1912; CR 1914; CR 1915; Plante Fébure, *West-Indië in het parlement*, p. 53; Hallewas, 'Gezondheidszorg', p. 43.
117 Verheggen, 'God is liefde', p. 58; Schalken, *Historische gids*, p. 58.
118 SN: F. P. Schuitemaker, 'De Lepra-bestrijding met Hydnocarpus-olie is geen neiuwe Engelsche vinding', *Suriname*, 12 April 1928; reprinted in *West-Indische Gids* 10 (1929), pp. 93–6.
119 CR 1916.
120 CR 1917; 'Ingekomen rapport voor het Koloniaal Verslag over 1916', GS 1243.
121 CR 1918; 'Ingekomen rapport voor het Koloniaal Verslag over 1918', GS 1246; CR 1919.

122 CR 1920, 1921.
123 CR 1922.
124 CR 1923.
125 CR 1925, pp. 19–20.
126 GS 1254.
127 CR 1926, p. 25; CR 1927, p. 26.
128 See L. C. van Panhuys, 'Zeer aanmoedigende uitkomsten van lepra-behandeling', *West-Indische Gids* 10 (1929), pp. 89–92.
129 League of Nations Leprosy Committee, League of Nations Room, United Nations, Geneva, 'Suriname' (hereafter LN),'Leprosy annual report 1928', pp. 2–3. (Documents courtesy of Jo Robertson.)
130 LN: 'Leprosy annual report 1929', pp. 2–5.
131 BP unnumbered.
132 BP T 197.
133 CR 1926, p. 25; CR 1927, p. 26.
134 CR 1910, Appendix 1 (Military Health Service).
135 CR 1926, p. 25; CR 1927, p. 26.
136 LN: 'Leprosy annual report 1929', p. 6.
137 S. J. Bueno de Mesquita, 'De behandeling van lepra in Bethesda met PROMIN en DIASONE', in *Vijftig jaren Protestantsche Melaatsen verpleging*, pp. 25–33. Financial difficulties in providing sulfone treatment: SB 51 (1948), p. 12.
138 J. Boom, 'Het levensverhaal van Humbert Willems. Opstand in de leprozerie Groot-Chatillon', *OSO* 22 (2003), pp. 117–23.

Conclusion

For Caribbean plantation economies to function and prosper, European colonizers needed Others – African slaves. In *Empire*, Michael Hardt and Toni Negri write about this production of Others, the creation of racial boundaries, and the dark Other as the negative component of European identity as well as the economic foundation of European economic systems. They identify contagious diseases as one of the most important threats to the boundaries between self and Other. For Hardt and Negri, 'The horror released by European conquest and colonization is a horror of unlimited content, flow and exchange – or really the horror of contagion, miscegenation and unbanded life. Hygiene requires protective barriers.'[1]

The modern history of leprosy cannot be understood without exploring this production of Others, which permeated colonial medicine in eighteenth-century Dutch Suriname and the Caribbean. Leprosy, as framed by Schilling and his contemporaries, was a disease of the Other, which had orginated in Africa through sloth, dirt, and lasciviousness, and had been transported in the bodies of the slaves to the New World. Leprosy was the epitome of the Other's difference, and had to be controlled to make the plantation economy work. Colonial medicine and public hygiene policies aimed to create a functional labour management and were needed to defend the boundaries between the Europeans and the Others. In the eighteenth century, leprosy was imagined as a continuous reminder of the permeability of racial boundaries, and as a colonial disease that threatened to be transmitted to the Netherlands, thus endangering not only the plantation economy in Suriname, but also the Dutch colonial empire as a whole. Managing leprosy meant colonizing the slaves' bodies and determining their options for movement, activity, and location. Compulsory segregation was hence not only the ultimate resort, but also a routine method of colonizing bodies in a policy that aimed to defend the boundary between the European and the Other.

In order to justify the control of the Other, at a conscious as well as an unconscious level, he or she had to be dehumanized. The presentation of slaves was suffused with what historian Keith Thomas calls 'the image of animality'. Slaves' lust and bestiality showed their quality of being subhuman. As animals, their only reason to exist was for the benefit of the summit of Creation – the white man.[2] Leprosy was racialized (a disease of the Other) and sexualized (transmitted by beast-like lust) and this introduced a distinct moral component to descriptions of the leprosy sufferer. Sufferers were black (and therefore inferior), ruled by lust (which showed how animal-like the sufferers were), and when the Other's lust was reciprocated by Europeans this threatened racial boundaries. The disease made the sufferer useless for the one purpose of his or her existence – performing labour.

In the stigmatization of the leprosy sufferer, the horrendous nature of the disease, the visible violation of purity, and the transgression of supernatural taboos, all played important roles. However, in the eighteenth-century framing of the disease and the justification of compulsory segregation, another significant factor was at play. For Europeans, leprosy represented the Other's most threatening face. The Other's very occurrence confirmed the bestiality of the leprosy sufferer, and it confirmed the bestiality of the group of people in which leprosy occurred the most. Therefore, colonial power relations were of central importance in the stigmatization of the leprosy sufferer.

In the eighteenth-century plantation economy, around leprosy a 'cultural meme', a complex of ideas and behaviours spreading from person to person in a specific culture, was created and maintained its power after the emancipation of the slaves and through the end of colonial rule. The meme surrounding leprosy was maintained because the Afro-Surinamese continued to be seen as a possible danger. While the Afro-Surinamese's importance as a labour force was eroded by the immigration of indentured labourers from British India and Java, their presence was still strong. As late as the interwar period, Afro-Surinamese unrest in the colony was manifest and the colonial state still felt compelled to suppress their religious rituals and practices (at least formally).

At the end of the nineteenth century, leprosy was reframed as an imperial danger on an international level. However, the history of the Surinamese colonial framing of leprosy had a distinct dynamic. Compulsory segregation was attached to the needs of modern public health policies. In the modernizing colonial state (the basis of leprosy politics), the need for the compulsory segregation of sufferers, and the presumed inferiority of the Afro-Surinamese Others, was never really challenged. Although the Dutch transferred some racial stigmatization of the leprosy sufferers to the British Indians, the Afro-Surinamese continued to constitute the majority of all the sufferers and received the greatest attention. Colonial medicine 'medicalized' leprosy policies, emphasizing the goal of treatment and the need for humane care. Christian churches fought over access and control of leprosy care, since this was of importance to their financing and the self-imaging of their missions, as well as the colonizing of the sufferers' souls. Colonial medicine did not always approve of the activities in Christian asylums, but these were minor differences when compared to the Europeans' shared general perspective, whether they were born in Suriname or sent from the Netherlands. From this perspective, leprosy sufferers had to be refashioned, disciplined, and made to accept religious ideas and practices other than their own. While colonial medicine was definitely interested in the belief systems of Afro-Surinamese sufferers in particular, there was never an idea of cultural, social, or racial equality. In their writings, colonial doctors, missionaries and others continued to preserve the 'cultural meme' of the age of slavery. Afro-Surinamese sufferers (approximately four out of five detected sufferers in Suriname in the first half of the twentieth century) were framed as lax in their health behaviour, superstitious in their practices, and unruly in their conduct. They were still Others who needed disciplining from above.

Reading the colonial sources produced by Europeans from a bottom-up perspective reveals another picture. It shows how the Afro-Surinamese rejected European views of leprosy as a typical African disease, and reframed it as a disease brought by Jews and/or connected to failed observance of the treef. Historical evidence on the extent to which leprosy sufferers in Suriname were stigmatized in their own social groups and families before the 1930s (when researchers began interviewing them) is not conclusive. However, a bottom-up perspective suggests that on the one hand, these people were left alone by their own groups, and were able to lead functional economic and social lives up to a point. On the other hand, once sufferers were segregated, the isolation from their own groups led to a problematic welcome if they returned. In other words, their segregated position was more the result of social standing and economic utility than a result of the disease itself. Furthermore, a bottom-up perspective makes it possible to penetrate colonial propaganda and acknowledge forms of resistance and contestation to reveal the diversity of the reactions of sufferers and their families. Here, infrapolitics of resentment, non-compliance, and fractious behaviour, and attempts to create a level of autonomy in everyday life is discernible.

Today, leprosy is a neglected tropical disease that has been conquered by tropical medicine according to the WHO. However, does Western management of health problems in tropical regions, or of other people of colour, continue to include traces of our colonial heritage? Recognizing and analysing these traces of colonial heritage and exploring the perspectives of other cultures are essential when investigating health and disease. This is significant since global migration movements make the permeability of boundaries and transmission of humans, and therefore diseases, more common then perhaps ever before in human history.

Notes

1 M. Hardt and T. Negri, *Empire* (Cambridge: Harvard University Press, 2000), pp. 135–6.
2 K. Thomas, *Man and the Natural World: Changing Attitudes in England 1500–1800* (London: Allen Lane, 1983), pp. 38, 44.

Sources and select bibliography

Primary sources, unpublished

Archives

AB:	Archive Bronbeek, Arnhem. (Documents courtesy of Leo van Bergen.)
AS:	National Archive, The Hague, Algemene secretarie Nederlandsche West-Indische bezittingen in Suriname 1830–1847. Inv. nr. 1.05.08.02.
BP:	Archive Bisdom of Paramaribo.
CE:	National Archive, The Hague, Handelingen en Bijlagen van de (Koloniale) Staten van Suriname. Inv. nr. 2.10.44.
CRO:	National Archives, Kew. Colonial Record Office
DS:	National Archive, Paramaribo, Districtscommissariaat Saramacca 1897–1955. Inv. nr. 1.24.01.
DTB:	National Archive, The Hague, Doop-, trouw- en begraafboeken van Suriname. Inv. nr. 1.05.11.16.
GS:	National Archive, Paramaribo, Gouvernementssecretarie Suriname 1829–1954. Inv. nr. 1.01.01.
GS II:	National Archive, The Hague, Gouverneur van Suriname 1885–1951. Inv. nr. 2.10.18.
LN:	League of Nations Leprosy Committee, League of Nations Room, United Nations, Geneva, 'Suriname'. (Documents courtesy of Jo Robertson.)
NA 2WC:	National Archive, The Hague, Tweede West-Indische Compagnie Archive. Inv. nr. 1.05.01.02.
ZZ:	Utrecht Historical Archive, Utrecht, Zeister Zendingsgenootschap.

Primary sources available online

CR: Colonial Reports. *Verslag van het beheer en den staat der Koloniën* (1849–1865); *Koloniaal Verslag* (1866–1923); *Verslag van bestuur en staat van Suriname* (1924–1930); *Surinaamsch Verslag* (1931–1950): www.statengeneraaldigitaal.nl/.
SA: Surinamese almanacs: www.dbnl.org/.
SG: Proceedings of the Dutch Parliament. *Handelingen Staten-Generaal*: www.statengeneraaldigitaal.nl/.
SN: Surinamese newspapers: www.delpher.nl/nl/kranten/.

Primary sources, printed

Journals
BB: *Brieven uit Bethesda* (Paramaribo: Protestantsche Melaatschen-inrichting Bethesda, 1934–1938).
GB: *Gouvernementsblad van de Kolonie Suriname* (Paramaribo: Gouvernement Suriname, 1816–1950).
GV: *De Godsdienstvriend* (Amsterdam: A. Schievenbus, 1818–1869).
SB: *Stemmen uit Bethesda* (Amsterdam: De Bussy, 1900–1951).

Books and articles
Beek, J. P. ter, 'Dissertatio medico-inauguralis de elephantiasi Surinamensi' (MD thesis, University of Leiden, 1841).
Benjamins, H. D., 'Treef', in H. D.Benjamins and J. F. Snelleman (eds.), *Encyclopaedie van Nederlandsch West-Indië* ('s-Gravenhage: Martinus Nijhoff, 1914–1917), pp. 685–7.
Benjamins, H. D., 'Treef en lepra in Suriname', *West-Indische Gids* 11 (1930), pp. 187–218.
Benjamins, H. D., and J. F. Snelleman (eds.), *Encyclopaedie van Nederlandsch West-Indië* ('s-Gravenhage: Martinus Nijhoff, 1914–1917).
Benoit, P. J., *Reis door Suriname. Beschrijving van de Nederlandse bezittingen in Guyana* (Zutphen: De Walburg Pers, 1980).
Bethesda. Een liefdewerk der protestanten in Suriname (Paramaribo: Protestantsche Vereeniging ter verpleging van Lepralijders in de Kolonie Suriname, 1902).

Sources and select bibliography

Blankensteijn, M. van, *Suriname* (Rotterdam: Nigh and Van Ditmar, 1923).
Blom, A., *Verhandeling over de landbouw in de colonie Suriname* (Amsterdam: J. W. Smit, 1787).
Bonne, C., 'De maatschappelijke beteekenis der Surinaamsche ziekten', *West-Indische Gids* 1 (1919), pp. 291–300.
Bosman, W., *Nauwkeurige beschrijving van de Guinese Goud-, Tand- en Slavenkust*. (Amsterdam: Isaac Stokmans, 1709).
Bosser, A., *Beknopte geschiedenis der katholieke missie in Suriname* (Gulpen: M. Alberts, 1884).
Broes van Dort, T., 'Een en ander over de lepra in Nederland en zijne koloniën', *Nederlandsch Tijdschrift voor Geneeskunde* 41 (1897), pp. 292–6, 384–91, 407–21, 650–1.
Broes van Dort, T., 'De internationale lepra-conferentie te Berlijn (11–16 Oct. 1897)', *Nederlandsch Tijdschrift voor Geneeskunde* 41 (1897), pp. 747–71, 810–15, 893–7, 937–42, 978–84.
Brons, J. C., *Het rijksdeel Suriname* (Haarlem: Bohn, 1952).
Büchner, W. F., *Geneeskundig handboek voor beginnende kunstoefenaren* (Amsterdam: H. J. Berntrop, 1839).
Canstatt, C., *Handbuch der medicinischen Klinik*, 3rd rev. ed., vol. 2 (Erlangen: Ferdinand Enke, 1855), pp. 1–15.
Carsten, B., 'Over de verspreidingswijze van lepra', *Nederlands Tijdschrift voor Geneeskunde* 11 (1867), pp. 481–5.
Cool, P., 'Regeeringsmaatregelen ter bestrijding der lepra in Suriname en Aruba', *Nederlandsch Tijdschrift voor Geneeskunde* 71 (1927), pp. 2453–6.
Coster, A. M., 'De boschnegers in de kolonie Suriname. Hun leven, zeden, en gewoonten', *Bijdragen tot de Taal-, Land- en Volkenkunde* 13 (1866), pp. 1–37.
Debien, G., 'Le marronage aux Antilles françaises au XVIIIe siècle', *Caribbean Studies* 6 (1966), pp. 3–43.
Deutschbein, L. L. A., 'De noma infantum' (MD thesis, University of Halle 1840).
Deutschbein, L. L. A., 'Report', *Tijdschrift voor de Wis- en Natuurkundige Wetenschappen* 5 (1852), pp. 100–5.
Dissel, J. A. van, 'Iets over de godsdienst der Javanen', *Tijdschrift voor Nederlandsch-Indië* 3rd series, 3 (1869), pp. 375–85.
Dresen, H. B., *Gouden jubilé van de R. K. leprozerie Sint Gerarus Majella Stichting* (Paramaribo: Sint Gerardus Majella Stichting, 1945).
Drognat Landré, C. L., *De besmettelijkheid der lepra arabum, bewezen door de geschiedenis dezer ziekte in Suriname* (Utrecht: J. L. Beijers, 1867).
Drognat Landré, C. L., *De la contagion seule cause de la propagation de la leprè* (Paris: Guillaume Baillière, 1869).

Essed, W. F. R., 'Eenige opmerkingen naar aanleiding van de artikelen over treef en lepra in dit tijdschrift verschenen', *West-Indische Gids* 12 (1931), pp. 257–67.
Fermin, P., *Traité des maladies les plus frequenter à Surinam, et des remedes les plus propres à les guérir* (Maastricht: Jacques Lekens, 1764).
Fermin, P., *Nieuwe algemeene beschryving van de colonie van Suriname* (Harlingen: Volkert van der Plaats, 1770).
Flu, P. C., *De filaria-ziekte in Suriname* (The Hague: Algemeene Landsdrukkerij, 1911).
Flu, P. C., 'Het een en ander over de besmetting met lepra', *Stemmen uit Bethesda* 29 (1924), pp. 56–7.
Flu, P. C., *Verslag van een studiereis naar Suriname (Nederlandsch Guyana.) Sept. 1927 – Dec. 1927* (Utrecht: Kemink, 1928).
Fock, L. C. E. E., *Natuur- en geneeskundig etymologisch woordenboek* (n.p.: J. Noorduyn, 1855).
Friedman, S., *Nederlandsch Oost- en West-Indië, volgens de nieuwste inrigting, met betrekking tot aardrijkskunde, statistieken, voortbrengselen, luchtgesteldheid, en vooral tot den gezondheidstoestand* (Amsterdam: J. C. A. Sülpke, 1861).
Fuchs, C. H., *Die krankhaften Veränderungen der Haut und ihrer Anhänge* (Göttingen: Dietrichsen Buchhandlung, 1840).
Gallandat, D. H., *Noodige onderrigtingen voor de slaafhandelaren* (Middelburg: Pieter Gilissen, 1769).
Gedenkboek van het koloniaal-militair invalidenhuis Bronbeek (Amsterdam: P. Gouda Quint, 1881).
Geschiedenis der kolonie van Suriname... Door een Gezelschap van geleerde joodsche mannen aldaar, vol. 2 (Amsterdam: Allart & Van der Plaats, 1791)
Gypser, K.-H. (ed.), *Herings Medizinische Schriften*, 3 vols. (Göttingen: Ulruch Burgdorf, 1988).
Hartsinck, J. J., *Beschrijving van Guiana of de Wilde Kust, in Zuid-Amerika*, 2 vols. (Amsterdam: Gerrit Tielenburg, 1770).
Hasewinkel, W., 'Jaarverslag van het Gesticht Bethesda, Suriname', *Berichten uit de Heidenwereld* 19 (1923), pp. 37–42.
Hasselaar, A. van, *Beschrijving der in de kolonie Suriname voorkomende elephantiasis en lepra (melaatschheid)* (Amsterdam: S. de Greber, 1835).
Herskovits, M. J., and F. S. Herskovits, *Suriname Folk-lore* (New York: Columbia University Press, 1936).
Hille, J., 'Ueber die Elephantiasis; nach eigenen Beobachtungen in West-Indien', *Wochenschrift für die gesammte Heilkunde* (1841), pp. 433–42.
Hillis, J. D., *Leprosy in British Guiana: An account of West Indian leprosy* (London: J. A. Churchill, 1881).

Houttuyn, M., *Handleiding tot de plant- en kruidkunde*, vol. 3 (Amsterdam: Lodewijk van Es, n.d.).
Huet, G. D. H., 'Een geval van lepra arabum. Lijkopening', *Nederlandsch Tijdschrift voor Geneeskunde* 12 (1868), pp. 113–20.
Jharap, E., *Vertrouwen in eigen kunnen* (The Hague: Amrit, 2007).
Jonge, B. C. de, *Herinneringen* (Groningen: Wolters–Noordhoff, 1968)
Kappler, A., *Surinam, sein Land, seine Natur, Bevölkerung und seine Kultur-Verhältnisse* (Stuttgart: J. G. Cotta, 1887).
Käyser, J. D., 'Beschouwingen naar aanleiding van het Verslag van de IIIde Internationale Lepraconferentie, gehouden te Straatsburg van 28–29 Juli 1923', *Geneeskundig Tijdschrift voor Nederlandsch-Indië* 65 (1925), pp. 716–50.
Kuhn, F. A., *Beschouwing van den toestand der Surinaamsche plantagieslaven. Eene oeconomisch-geneeskundige bijdrage tot verbetering deszelven* (Amsterdam: C. G. Sulpke, 1828).
Kuhn, F. A., 'Over de elephantiasis te Suriname', *Hippocrates* 7 (1828), pp. 12–28.
Kuijs, A. P., and G. J. Rijnders, *Waarneming eener Elephantiasis aan het linkerbeen* (Amsterdam: Ten Brink & De Vries, 1820).
Lampe, P. H. J., 'Enkele opmerkingen over den sociaal-hygienischen toestand en de geneeskundige verzorging van Suriname', *West-Indische Gids* 8 (1927), pp. 249–76.
Lampe, P. H. J., 'Sociaal-hygiënische beschouwingen' (Kon. Vereeniging Koloniaal Instituut, Mededeeling no. XXIII, 1927).
Lampe, P. H. J., 'Het Surinaamsche treefgeloof. Een volksgeloof betreffende het ontstaan van de melaatschheid', *West-Indische Gids* 10 (1929), pp. 545–68.
Lampe, P. H. J., and R. Boenjamin, 'Social intercourse with lepers and the subsequent development of manifest leprosy', *Documenta Neerlandica et Indonesica de morbis tropicis* 1 (1949), pp. 289–346.
Lampe, P. H. J., and C. Simons, 'Lepra in Suriname', *Nederlandsch Tijdschrift voor Geneeskunde* 73 (1929), pp. 4903–15.
Lampe, R., *'Ik werd militair'* (n.p., 1947).
Landré, C., 'Bijdragen tot de kennis der ziekten van de negers in de kolonie Suriname', *Nieuw Praktisch Tijdschrift voor Geneeskunde in al haren omvang*, 31 (Nieuwe reeks 4, 1852), pp. 496–7.
Landré, C. H., 'Naschrift bij P. Duchassaing, Over de Elephantiasis Arabum in West-Indië', *West-Indië. Bijdragen tot de bevordering van de kennis der Nederlandsch West-Indische koloniën*, 2 (1858), pp. 222–33.
Landré, C., 'Sur la contagion de la lepre', in F. J. van Leent A. A. G. Guye, de Perrot, and J. Zeeman (eds.), *Congrès international de medicine des colonies, Amsterdam, Septembre 1883* (Amsterdam: F. van Rossen, 1884), pp. 277–9.

Landré, C., *Over de oorzaken der verbreiding van de lepra. Een waarschuwend woord hoofdzakelijk gericht tot de bewoners van Suriname* (The Hague: Martinus Nijhoff, 1889).
Langen, C. D. de, and A. Lichtenstein, *Leerboek der tropische geneeskunde*, 2nd rev. ed. (Weltevreden: G. Kolff, 1928).
Leent, F. J. van, 'Review of Landré, *Besmettelijkheid*', *Geneeskundig Tijdschrift voor de Zeemagt* 7 (1869), pp. 63–75.
Lemmens, A. F., *Bijdragen tot de Kennis van de Kolonie Suriname. Tijdvak 1816–1822* (Geografisch en Planologisch Instituut VU Amsterdam, 1982).
Lens, T., 'Lepra in Suriname', *Elsevier's Geïllustreerd Maandschrift* 10 (1895), pp. 521–52.
Lichtveld, L., 'Op zoek naar de spin', *West-Indische Gids* 12 (1931), pp. 209–30.
Lier, W. F. van, 'Aanteekeningen over het geestelijk leven en de samenleving der Djoeka's (Aukaner Boschnegers) in Suriname', *Bijdragen tot de Taal-, land- en Volkenkunde* 99 (1940) pp. 130–294, on pp. 273–6.
Ludwig, J. F., *Neueste Nachrichten von Surinam* (Jena: Akademischen Buchhandlung, 1789).
Maijer, L. T., *De Javaan als doekoen. Een ethnografische bijdrage* (Weltevreden: G. Wolff, 1918).
Manson, P., *Tropical Diseases: A Manual of the Diseases of Warm Climates*, rev. ed. (London: Cassell, 1903).
May, T., 'De lepra, haar voorkomen, verspreiding en bestrijding, in 't bijzonder in Suriname', I, *West-Indische Gids* 8 (1927), pp. 547–56, II *West-Indische Gids* 9 (1928), pp. 17–31.
'Mededeelingen nopens de lepra in onze West-Indische bezittingen', *Nieuw Praktisch Tijdschrift voor de Geneeskunde* 28, Nieuwe reeks 1 (1849), pp. 546–68, 761–70.
Miitheilungen und Verhandlungen der internationalen wissenschaftlichen Lepra-Conferenz zu Berlin, 3 vols. (Berlin: August Hirschwald, 1897).
Oudermeulen, B. van der. 'Iets tot voordeel der deelgenooten van de Oost-Indische Compagnie en tot nut van ieder ingezetenen van dit gemenebest kan strekken', in D. van Hogendorp (ed.), *Stukken, raakende den tegenwoordige toestand der Bataafsche bezittingen in Oost-Indië en de handel op derzelve* (The Hague: J. C. Leeuwesteyn, 1801), pp. 327–38, on pp. 327–8.
Panhuys, L. C. van, 'About the "trefe" superstition in the colony of Surinam', *Janus* 28 (1924).
Panhuys, L. C. van, 'Zeer aanmoedigende uitkomsten van lepra-behandeling', *West-Indische Gids* 10 (1929), pp. 89–92.
Panhuys, L. C. van, 'De opoffering van een R. K. priester in Suriname', *West-Indische Gids* 18 (1937), pp. 201–6.

Penard, F. P., and A. Penard, 'Surinaamsch bijgeloof'. Iets over Winti en andere natuurbegrippen', *Bijdragen tot de Taal-, Land- en Volkenkunde van Nederlandsch-Indië* 67 (1913), pp. 157–89.

Peschuël-Loesche, P., *Volkskunde von Loango* (Stuttgart: Stecker and Schröder, 1907).

Plante Fébure, J. M., *West-Indië in het parlement 1897–1917, Bijdrage tot Nederland's koloniaal-politieke geschiedenis* ('s-Gravenhage, Martinus Nijhoff, 1918).

Reddingius, R. A., 'Review of Landré, Oorzaken', *Nederlandsch tijdschrift voor Geneeskunde* 33 (1889), p. 586.

Reijnders, B. J. C., 'Het werk van de Evangelische Broedergemeente in Suriname', *West-Indische Gids* 28 (1947), pp. 300–11.

'Report Commissie van Geneeskundig Onderzoek en Toevoorzigt', *Nieuw Praktisch Tijdschrift voor de Geneeskunde in al haren omvang* 1 (1849), pp. 554–65.

Rodschied, E. K., *Medizinische und Chirurgische Bemerkungen über das Klima, die Lebensweise und Krankheiten der Einwohner der Holländischen Kolonie Rio Essequibo* (Frankfurt: Jaegerschen Buchhandlung, 1796).

Rogers, L., *Happy Toil: Fifty-five Years of Tropical Medicine* (London: Frederick Muller, 1950).

Rolander, D., 'Journal', in L. Hansen, D. Goodall, and J. Dobreff (eds.), *The Linnaeus Apostles: Global Science and Adventure*, vol. 3, bd. 3 (London: IK Foundation, 2008), pp. 1217–564.

Sanders, C. H., *De melaatschheid* (Groningen: R. J. Schierbeek, 1867).

Schaick, C. van, *De manja, Familietafereel uit het Surinaamsche volksleven* (Arnhem: D. A. Thieme, 1866).

Schilling, G. W., 'Dissertatio medica inauguralis de lepra' (MD thesis, University of Utrecht, 1769).

Schilling, G. W., *Diatribe de morbo in Europa pene ignoto, quem America vocant Jaws* (Utrecht: J. C. ten Bosch, 1770).

Schilling, G. W., *Geneeskundige verhandeling van eene in Europa byna onbekende ziekte, bij de Amerikanen JAWS genoemd* (Middelburg: Christiaan Bohemer, 1770).

Schilling, G. W., *Verhandeling over de melaatsheid* (Utrecht: J. C. ten Bosch, 1771).

Schilling, G. W., 'Animadversiones in Ouseelianam et additamenta ad suam de lepra dissertationem', in J. D. Hahn (ed.), *De lepra commentationes* (Leiden: Abr. van Paddenburg, 1778), pp. 119–203.

Schneevoogt, V., 'Verslag op het rapport van den heer Ooijkaas, omtrent het lepreuzen etablissement Batavia, in de kolonie Suriname', *Verslagen en Mededeelingen der Koninklijke Akademie van Wetenschappen* 2 (1854), pp. 381–8.

Schönfeld. K. D., 'Verhandeling over de lepra in 't algemeen, en de elephantiasis tuberculosa in 't bijzonder' (MD-thesis, University Groningen 1857).
Schuitemaker, F. P., *De lepra en de gouvernementsleproserie in Suriname* (Amsterdam: J. H. de Bussy, 1915).
Schuitemaker, F. P., 'De Lepra-bestrijding met Hydnocarpus-olie is geen neiuwe Engelsche vinding', Suriname 12 April 1928; reprinted in *West-Indische Gids* 10 (1929), pp. 93-6.
Schuitemaker, F. P., 'Bezoek aan St. Louis, het leprozeneiland in de Marowyne; melaatschen etablissement van de Fransche strafkolonie St Laurent, Fransch-Guyana', *De West-Indische Gids* 11 (1930), pp. 177-86.
Schweigman, F., *Twee Missionarissen onder de Melaatschen en Indianen van Suriname* (Roermond: J. J. Remen, 1894), pp. 1-173.
Schweigman, F., *Aan de leden van het Hofbauer-liefdewerk: Pater Donders* (Amsterdam: Borg, 1900).
Simons, R. D. G. P., 'De maatschappelijke beteekenis der Surinaamsche ziekten', *West-Indische Gids* 14 (1933), pp. 429-39.
Simons, R. D. G. P., *Lepra. De maligne contagieuze Morbus Hansen en de benigne niet-contagieuze Hanseniden* (Amsterdam: Van Holkema and Warendorf, 1948).
Simons, R. D. G. P., *Lepra. De lepra-bestrijding in Suriname en de noodzakelijkheid harer reorganisatie* (Amsterdam: Scheltema and Holkema, 1950).
Simons, R. D. G. P., *Bijgeloof en lepra in de Atlantische Negerzônes* (Paramaribo: Radhakishun, 1959).
Smidt, J. T. (ed.), *Plakkaten, ordonnantiën en andere wetten, uitgevaardigd in Suriname 1667-1816* (Amsterdam: Emmering, 1973).
Stedman, J. G., *Narrative of a Five Years' Expedition against the Revolted Negroes of Surinam, in Guiana on the Wild Coast of South America from the Years 1772 to 1777* (London: J. Johnson & J. Edwards, 1796; repr. Amherst: University of Massachusetts Press, 1972).
Stern, I., *Sternen-Saat* (n.p.: Curitiba, 2010).
Surinaamsche Almanach, op het jaar onzer Heere Jesu Christi Anno 1789 (Paramaribo: W. H. Poppelman, 1789).
Teenstra, M. D., *De landbouw in de kolonie Suriname* (Groningen: Eekhoff, 1835).
Verslagen over de lepra te Suriname (Amsterdam: G. M. P. Landonck, 1851).
Veertig jaren Protestantsche Melaatschen verpleging "Bethesda" in Suriname (Paramaribo: Protestantsche Vereeniging, 1939).
Verheggen, A., *God is liefde, Veertig jaren melaatschenverpleging 1895-1935* (Maastricht: Gebr. Van Aelst, 1935).

Verslag der herdenking van het 25-jarig bestaan van "Bethesda" op 15 Mei 1924 (Paramaribo: Bethesda, 1924).
Veth, P. J., *Java. Geographisch, ethnologisch, historisch*, 2nd rev. ed., vol. 4 (Haarlem: De Erven Bohn, 1907).
Vijftig jaren Protestantse Melaatsen verpleging "Bethesda" in Suriname 1899– 1949 (Paramaribo: Protestantsche Vereeniging, 1949).
Vinkhuijzen, *De melaatschheid, vooral met betrekking tot hare oorzaken en verhouding in de maatschappij* ('s-Gravenhage: De Gebroeders Van Cleef, 1868).
Virchow, R., *Die krankhaften Geschwülste. Dreissig Vorlesungen*, I (Berlin: August Hirschwald).
Weigel, M., *Brieven uit Bethesda* (Paramaribo: Nieuw-Bethesda, 1934).
Wesenhagen, A. C., *Suriname, Iets over land en volk* (Amsterdam: J. H. de Bussy, 1896).
Wolff, J. W., 'Het lepra-probleem in Suriname', *Geneeskundig Tijdschrift voor Nederlandsch-Indië* 65 (1925), pp. 572–87.
Woensel, P. van, 'West-Indische fragmenten', in A. Hanou (ed.), *De lantaarn* (Amsterdam: Athenaeum-Polak & Van Gennep, 2002).
Wolbers, J., *Geschiedenis van Suriname* (Amsterdam: De Hoogh, 1861).
Wright, H. P., *Leprosy, an Imperial Danger* (London: Churchill, 1889).

Secondary sources, published books and articles

Abbenhuis, M. F., 'De katholieke kerk in Suriname', *Vox Guyanae* 2 (1956), pp. 117–44.
Alveiz Moreira, T. M., and C. M. Varkevisser, *Gender, Leprosy and Leprosy Control: A Case Study in Rio de Janeiro State, Brazil* (Amsterdam: KIT, 1999).
Andel, T. van, 'The reinvention of household medicine by enslaved Africans in Suriname', *Social History of Medicine* 28 (2015), doi: 10.1093/shm/hkv014.
Andel, T. van, and S. Ruysschaert, *Medicinale en religieuze planten van Suriname* (Amsterdam: KIT, 2011).
Andel, T. van, and S. Ruysschaert, 'What makes a plant magical? Symbolism and sacred herbs in Afro-Surinamese Winti rituals', in R. Voeks and J. Rashford (eds.), *African Ethnobotany in the Americas* (New York: Springer, 2011), pp. 247–84.
Anderson, W., 'Leprosy and citizenship', *Positions* 6 (1998), pp. 707–30.
Anderson, W., *The Cultivation of Whiteness: Science, Health, and Racial Destiny in Australia* (New York: Basic Books, 2003).
Anderson, W., *Colonial Pathologies: American Tropical Medicine, Race and Hygiene in the Philippines* (Durham, NC: Duke University Press, 2006).

Arnold, D., 'Introduction: Disease, medicine and empire', in D. Arnold (ed.), *Imperial Medicine and Indigenous Society* (Manchester: Manchester University Press, 1988), pp. 1–26.

Arnold, D., *Colonizing the Body: State Medicine and Epidemic Disease in Nineteenth-Century India* (Delhi: Oxford University Press, 1993).

Au, S., *Mixed medicines: Health and Culture in French Colonial Cambodia* (Chicago, IL: The University of Chicago Press, 2011).

Bankole, K., *Slavery and Medicine: Enslavement and Medical Practices in Antebellum Louisiana* (New York: Garland, 1998).

Bashford, A., *Imperial Hygiene: A Critical History of Colonialism, Nationalism and Public Health* (Basingstoke: Palgrave Macmillan, 2004).

Beeldsnijder, R. O., '"Om werk van jullie te hebben". Plantageslaven in Suriname, 1730–1750' (PhD thesis, Leiden University, 1994).

Bergen, L. van, and S. Snelders (eds.), 'Van piratendokters tot wetenschappelijke instituten. Drie eeuwen Nederlandse en Belgische tropische geneeskunde', *Studium* 2 (2009), pp. 53–129.

Berlin, I., and P. D. Morgan (eds.), *The Slaves' Economy: Independent Production by Slaves in the Americas* (London: Routledge, 1995).

Bewell, A., *Romanticism and Colonial Disease* (Baltimore, MD: The Johns Hopkins University Press, 1999).

Bhagwanbali, R., *De nieuwe awatar van slavernij. Hindoestaanse migranten onder het indentured labour system naar Suriname, 1873–1916* (The Hague: Amrit, 2010).

Bijker, K., 'Power, prayer and colonial pacification: The Roman Catholic mission in nineteenth century Surinam', in M. Bax and A. Koster (eds.), *Power and Prayer: Religious and Political Processes in Past and Present* (Amsterdam: VU University Press, 1993), pp. 57–79.

Blom, A., 'Angst voor lepra', *OSO* 22 (2003), pp. 90–8.

Blom, A., 'Lepra in Suriname'. *MensenBeelden* 6 (2004), pp. 28–31.

Blom, J. C. H., and J. Talsma, *De verzuiling voorbij. Godsdienst, stand en natie in de lange negentiende eeuw* (Amsterdam: Het Spinhuis, 2000).

Boom, J., 'Het levensverhaal van Humbert Willems. Opstand in de leprozerie Groot-Chatillon', *OSO* 22 (2003), pp. 117–23.

Brouwers, J., *Na de drie begijnen ging het verder. Geschiedenis van de Congregratie van de Zusters van Liefde van Onze Lieve Vrouw, Moeder van Barmhartigheid* ('s-Hertogenbosch: Congregratie Zusters van Liefde, 2000), pp. 98–120.

Buckingham, J., 'The "morbid mark": The place of the leprosy sufferer in nineteenth century Hindu law', *South Asia: Journal of South Asian Studies* 20 (1997), pp. 57–80.

Buckingham, J., *Leprosy in Colonial South India: Medicine and Confinement* (Basingstoke: Palgrave, 2002).

Buckingham, J., 'The inclusivity of exclusion: Isolation and community among leprosy-affected people in the South Pacific', *Health and History* 13 (2011), pp. 65–83.

Buddingh', H., *De geschiedenis van Suriname* (Amsterdam: Nieuw Amsterdam, 2012).

Burns, S. L., 'From "leper villages" to leprosaria: Public health, nationalism and the culture of exclusion in Japan', in C. Strange and A. Bashford (eds)., *Isolation: Places and Practices of Exclusion* (London: Routledge, 2003), pp. 104–18.

Cohen, R., *Jews in Another Environment: Surinam in the Second Half of the Eighteenth Century* (Leiden: E. J. Brill, 1991).

Cook, H. J., *Matters of Exchange: Commerce, Medicine, and Science in the Dutch Golden Age* (New Haven, CT: Yale University Press, 2007).

Crosby, Jr., A. W., *The Columbian Exchange: Biological and Cultural Consequences of 1492* (Westport, CT: Greenwood, 1972).

Dankelman, J. L. F., *Peerke Donders. Schering en inslag van zijn leven* (Hilversum: Gooi en Sticht, 1982).

Davis, G., *Holy Man: Father Damien of Molokai* (Honolulu: University of Hawai'i Press, 1973).

Davis, N. Z., 'Physicians, healers, and their remedies in colonial Suriname', *Canadian Bulletin of Medical History/Bulletin canadien d'histoire de la médecine* 33 (2016), pp. 3–34.

Demaitre, L., *Leprosy in Premodern Medicine: A Malady of the Whole Body* (Baltimore, MD: The Johns Hopkins University Press, 2007).

Donselaar, J., van, *Woordenboek van het Nederlands in Suriname van 1667 tot 1867* (Amsterdam: Meertens Insitituut, 2013).

Eckart, W. U., *Medizin und Kolonialimperialismus: Deutschland 1884–1945* (Paderborn: Schöningh, 1997).

Edmond, R., *Leprosy and Empire: A Medical and Cultural History* (Cambridge: Cambridge University Press, 2006).

Eerenbeemt, A. J. J. M., *De missie-actie in Nederland (±1600–1940)* (Nijmegen: J. J. Berkhout, 1946).

Einaar, J. F. E., *Bijdrage tot de kennis van het Engelsch tussenbestuur van Suriname 1804–1816* (Leiden: M. Dubbeldeman, 1934).

Eyk, P. J. van, 'Oorlogsjaren in Suriname. Nederlands koloniaal beleid binnen Amerikaanse marges', *OSO* 14 (1995), pp. 148–57.

Farley, J., *Bilharzia: A History of Imperial Tropical Medicine* (Cambridge: Cambridge University Press, 1991).

Fermor, P. L., *The Traveller's Tree: A Journey Through the Caribbean Islands* (New York: New York Review Books, 2011).

Fett, S. M., *Working Cures: Healing, Health, and Power on Southern Slave Plantations* (Chapel Hill: The University of North Carolina Press, 2002).

Foucault, M., *Madness and Civilization: A History of Insanity in the Age of Reason* (London: Routledge, 2001).
Fox-Genovese, E., and E. D. Genovese, *The Mind of the Master Class: History and Faith in the Southern Slaveholders' Worldview* (Cambridge: Cambridge University Press, 2005).
Garraway, D., *The Libertine Colony: Creolization in the French Caribbean* (Durham: Duke University Press, 2005).
Genovese, E. D., *Roll, Jordan, Roll: The World the Slaves Made* (New York: Pantheon, 1974).
Gómez Zuluaga, P. F., 'Bodies of Encounter: Health, Illness and Death in the Early Modern African-Spanish Caribbean' (PhD thesis, Vanderbilt University, 2010).
Gómez, P. F., 'The circulation of bodily knowledge in the seventeenth-century black Spanish Caribbean', *Social History of Medicine* 26 (2013), pp. 383–402.
Gould, T., *Don't Fence Me In: Leprosy in Modern Times* (London: Bloomsbury, 2005).
Govers, N., *Leven van den eerbiedwaardigen Petrus Donders C.ss.R. Apostel der indianen en melaatsen in Suriname* (Heerlen: Joh. Roosenboom, 1946).
Gussow, Z., *Leprosy, Racism, and Public Health: Social Policy in Chronic Disease Control* (Boulder, CO: Westview Press, 1989).
Hallewas, G.-J., 'De gezondheidszorg in Suriname' (PhD thesis, Groningen University, 1981).
Handler, J., 'Diseases and medical disabilities of enslaved Barbadians, from the seventeenth century to around 1838', *The Journal of Caribbean History*, 40 (2006), pp. 1–38, 177–214.
Hardt, M., and T. Negri, *Empire* (Cambridge, MA: Harvard University Press, 2000).
Harrison, M., 'The tender frame of man: Disease, climate, and racial differences in India and the West Indies', *Bulletin of the History of Medicine* 70 (1996), pp. 68–93.
Harrison, M., *Medicine in an Age of Commerce and Empire: Britain and Its Tropical Colonies 1660–1830* (Oxford: Oxford University Press, 2010).
Hassankhan, M. S., 'De immigratie en haar gevolgen voor de Surinaamse samenleving', in L. Gobardhan-Rambocus and M. S. Hassankhan (eds.), *Immigratie en ontwikkeling. Emancipatieproces van contractanten* (Paramaribo: Anton de Kom Universiteit, 1993), pp. 11–35.
Hassankhan, M. S., B. V. Lal, and D. Munro (eds.), *Resistance and Indian Indentured Experience: Comparative Perspectives* (New Delhi: Manohar, 2014).
Haynes, D. M., *Imperial Medicine: Patrick Manson and the Conquest of Tropical Disease* (Philadelphia, PA: University of Pennsylvania Press, 2001).

Helmers, H., *Een groot Nederlander in Suriname. Leven en werken van den eerbiedw. Dienaar Gods Petrus Donders* (Tilburg: Henri Bergmans, 1946).
Hesselink, L., *Healers on the Colonial Market: Native Doctors and Midwives in the Dutch East Indies* (Leiden: KITLV Press, 2011).
Heuman, G., *The Caribbean* (London: Hodder Arnold, 2006).
Hinte-Rustwijk, D. van, and G. van Steenderen-Rustwijk, 'Van bedrijfsschade tot verzuilde paria', *OSO* 22 (2003), pp. 10–20.
Hoefte, R., *De betovering verbroken. De migratie van Javanen naar Suriname en het rapport-Van Vleuten (1909)* (Dordrecht: Foris, 1990).
Hoefte, R., *In Place of Slavery: A Social History of British Indian and Javanese Laborers in Suriname* (Gainesville, FL: University Press of Florida, 1998).
Hoefte, R., *Suriname in the Long Twentieth Century: Domination, Contestation, Globalization* (New York: Palgrave MacMillan, 2014).
Hoogbergen, W., and H. Ramsoedh (eds.), 'Lepra in Suriname', *OSO*, 22 (2003), pp. 1–123.
Hove, O. ten, '19e eeuws bevolkingsonderzoek naar lepra in Suriname', *OSO* 22 (2003), pp. 34–49.
Hulme, P., *Colonial Encounters: Europe and the Native Caribbean, 1492–1797* (London: Methuen, 1986).
Hyam, R., *Empire and Sexuality: The British Experience* (Manchester: Manchester University Press, 1990).
Ingliss, K. I., *Disease and Displacement in Nineteenth-Century Hawai'I* (Honolulu: University of Hawai'i Press, 2013).
Ismael, J., 'De immigratie van Indonesiërs in Suriname' (PhD thesis, Leiden, 1949).
Jensen, N. T., *For the Health of the Enslaved: Slaves, Medicine and Power in the Danish West Indies, 1803–1848* (Copenhagen: Museum Tusculaneum Press, 2012).
Kakar, S., 'Leprosy in India: The intervention of oral history', *Oral History* 23 (1995), pp. 37–45.
Kakar, S., 'Leprosy in British India, 1860–1940: Colonial politics and missionary medicine', *Medical History* 40 (1996), pp. 215–30.
Karbaat, J., '200 jaar militair hospitaal in Paramaribo (1760–1960)', *Nederlands Militair Geneeskundig Tijdschrift* 13 (1960), pp. 355–64.
Karbaat, J., 'Sociaal-geneeskundige beschouwingen over de personeelsleden van de troepenmacht in Suriname en hun gezinnen' (MD-thesis, Leiden University, 1963).
Karbaat, J., 'De historie van de militair-geneeskundige dienst in Suriname', *Nederlands Militair Geneeskundig Tijdschrift* 17 (1964), pp. 275–7.
Kerkhoff, A. H. M., 'The organization of the military and civil military service in the nineteenth century', in G. M. van Heteren. A. de Knecht-van Eekelen and

M. J. D. Poulissen (eds.), *Dutch Medicine in the Malay Archipelago 1816-1942* (Amsterdam: Rodopi, 1989), pp. 9-24.

Ketting, 'Bijdrage tot de geschiedenis van de lepra in Nederland' (MD-thesis, University of Amsterdam, 1922).

Ki Che Leung, A., *Leprosy in China: A History* (New York: Columbia University Press, 2008).

Kiple, K. F., *The Caribbean Slave: A Biological History* (Cambridge: Cambridge University Press, 1984).

Kiple, K. F., and Kriemhild Coneé Ornelas, 'Race, war and tropical medicine in the eighteenth-century Caribbean', in D. Arnold (ed.), *Warm Climates and Western Medicine: The Emergence of Tropical Medicine, 1500-1900* (Amsterdam: Rodopi, 1996), pp. 65-79.

Kleijntjes, J., 'Mgr. Grooff, apostolisch vicaris van Batavia', *Bijdragen voor de Geschiedenis van het Bisdom van Haarlem* 47 (1931), pp. 373-468.

Kleinman, A., *Patients and Healers in the Context of Culture: An Exploration of the Borderland between Anthropology, Medicine, and Psychiatry* (Berkeley, CA: University of California Press, 1980).

Klerk, C. J. M., *De immigratie van Hindoestanen in Suriname* (Amsterdam: Urbi et Orbi, 1953).

Klinkers, E., 'De bannelingen van Batavia. Lepra-bestrijding gedurende de negentiende eeuw in koloniaal Suriname', *OSO* 22 (2003), pp. 50-61.

Klinkers, E., *De geschiedenis van de politie in Suriname, 1863-1975. Van koloniale tot nationale ordehandhaving* (Amsterdam: Boom, 2010).

Kronenburg, J. A. F., *De eerbiedw, dienaar Gods Petrus Donders C.ss.R. Nieuwe levensbeschrijving* (Tilburg: W. Bergmans, 1925).

Kuyp, E, van der, 'Surinaamse medische en paramedische kroniek 1494-1899', *Surinaams Medisch Bulletin* 9 (1985), pp. 1-67.

Kuyp, E. van der, 'De geschiedenis van lepra in Suriname tot 1971', *Surinaams Medisch Bulletin* 14 (1999), 1, pp. 43-64; 2, pp. 36-55.

Laguerre, M., *Afro-Caribbean Folk Medicine* (South Hadley, MA: Begrin & Garvey, 1987).

Lamur, H. E., 'The Demographic Evolution of Surinam 1920-1970: A Socio-Demographic Analysis' (PhD thesis, University of Amsterdam, 1973).

Laurence, B. R., '"Barbadoes leg": Filariasis in Barbados, 1625-1900', *Medical History* 33 (1989), pp. 480-8.

Lenders, M., *Strijders voor het lam. Leven en werk van Herrnhutter broeders en zusters in Suriname, 1735-1900* (Leiden: KITLV Press, 1996).

Lier, R. van, *Frontier Society: A Social Analysis of the History of Surinam* (The Hague: Martinus Nijhoff, 1971).

Lijphart, A., *The Politics of Accommodation: Pluralism and Democracy in the Netherlands* (Berkeley: University of California Press, 1975).

Lindeboom, G. A. *Dutch Medical Biography: A Biographical Dictionary of Dutch Pphysicians and Surgeons 1475–1975* (Amsterdam: Rodopi, 1984).
MacLeod, R., 'Introduction', in R. MacLeod and M. Lewis (eds.), *Disease, Medicine, and Empire: Perspectives on Western Medicine and the Experience of European Expansion* (London and New York: Routledge, 1988), pp. 1–18.
MacLeod, R. (ed.), 'Nature and empire: Science and the colonial enterprise', *Osiris* 15 (2000), pp. 1–317.
McCollin, D., 'Chacachacare: The island of lepers, 1922–1979', in C. Bonfield, J. Reinarz, and T. Huguet-Termes, *Hospitals and Communities, 1100–1960* (Oxford: Peter Lang, 2013), pp. 263–90.
McNeill, J. R., *Mosquito Empires: Ecology and War in the Greater Caribbean, 1629–1914* (Cambridge: Cambridge University Press, 2010).
Menke, H., interview with Paul Niemel, 'Lepra in Suriname: van segregratie naar integratie', *OSO* 22 (2003), pp. 21–33.
Menke, H., S. Snelders, and T. Pieters, 'Omgang met lepra in 'de West' in de negentiende eeuw. Tegendraadse maar betekenisvolle geluiden vanuit Suriname', *Studium* 2 (2009), pp. 65–77.
Mintz, S. W., and R. Price, *The Birth of Afro-American Culture: An Anthropological Perspective* (Boston, MA: Beacon Press, 1992).
Murto, C., C. Kaplan, L. Ariza, K. Schwarz, C. H. Alencar, L. M. M. da Costa, and J. Heukelbach, 'Factors associated with migration in individuals affected by leprosy, Maranhão, Brazil: An exploratory cross-sectional study', *Journal of Tropical Medicine* 2013: doi:10.1155/2013/495076.
Navon, L., 'Beggars, metaphors, and stigma: A missing link in the social history of leprosy', *Social History of Medicine* 11 (1998), pp. 89–106.
Obregon, D., 'Building national medicine: Leprosy and power in Colombia, 1870–1910', *Social History of Medicine* 15 (2002), pp. 89–108.
Oostindie, G., *Roosenburg en Mon Bijou. Twee Surinaamse plantages, 1720–1870* (Dordrecht: Fortis, 1989).
Oudschans Dentz, F., 'De. Constantin Hering en Christiaan Johannes Hering', *West-Indische Gids* 12 (1931), pp. 147–60.
Oudschans Dentz, F., *De kolonisatie van de Portugeesch Joodsche natie in Suriname en de geschiedenis van de Joden Savanne* (Amsterdam: S. Emmering, 1975).
Pandya, S., 'The first international leprosy conference, Berlin 1897', *Manguinhos* 10, suppl. 1 (2003), pp. 161–77.
Panhuys, L. C. van, 'De Gouverneur-Generaal Willem Benjamin van Panhuys', *De West-Indische Gids* 6 (1925), pp. 291–320.
Peckham, R., and D. M. Pomfret (eds.), *Imperial Contagions: Medicine, Hygiene, and Cultures of Planning in Asia* (Hong Kong: Hong Kong University Press, 2013).

Pluchon, P., *Nègres et juifs au XVIIIe siècle. Le Racism au siècle des Lumières* (Paris: Tallandier, 1984).
Postma, J., 'De leproserie Bethesda tussen 1897 en 1928', *OSO* 22 (2003), pp. 69–81.
Quinian, S., 'Colonial encounters: Colonial bodies, hygiene and abolitionist policies in eighteenth-century France', *History Workshop Journal* 42 (1996), pp. 107–26.
Ramsoedh, H., 'Suriname 1933–1944. Koloniale politiek en beleid onder gouverneur Kielstra' (PhD thesis, Utrecht University, 1990).
Ramsoedh, H., 'Rumcola en Yankeedollars', *OSO* 14 (1995), pp. 134–47.
Robertson, J., 'In a State of Corruption: Loathsome Disease and the Body Politic' (PhD thesis, University of Queensland, 1999), http://espace.library.uq.edu.au/view/UQ:193252/the13742.pdf. Accessed 21 October 2014.
Robertson, J., 'Leprosy and the elusive M. Leprae: Colonial and imperial medical exchanges in the nineteenth century', *Manguinhos*, 10; suppl.1 (2003), pp. 13–40.
Robertson, J., 'The leprosy asylum in India, 1886–1947', *Journal for the History of Medicine and Allied Sciences* 64 (2009), pp. 474–517.
Rogozinski, J., *A Brief History of the Caribbean: From the Arawak and the Carib to the Present* (New York: Facts on File, 1999).
Rolander, D., 'Journal', in L. Hansen, D. Goodall, and J. Dobreff. (eds.), *The Linnaeus Apostles: Global Science and Adventure*, vol. 3, bd. 3 (London: IK Foundation, 2008), pp. 1217–564.
Savitt, T. L., *Medicine and Slavery: The Diseases and Health Care of Blacks in Antebellum Virginia* (Urbana, IL: University of Illinois Press, 1978).
Schalken, A. C., 'Historische gids bestaande uit chronologische lijst naamlijsten varia registers. 300 jaar R.K.-gemeente in Suriname 1683–1983' (Paramaribo, n.p., 1985).
Schalkwijk, J. M. W., *The Colonial State in the Caribbean: Structural Analysis and Changing Elite Networks in Suriname, 1650–1920* (The Hague: Amrit, 2011).
Schalkwijk, M., 'The plantation economy and the capitalist mode of production', in M. Schalkwijk and S. Small (eds.), *New Perspectives on Slavery and Colonialism in the Caribbean* (The Hague: Hamrit/Ninsee, 2012), pp. 14–40.
Schiebinger, L., 'The anatomy of difference: Race and sex in eighteenth-century science', *Eighteenth-Century Studies* 23 (1989/1990), pp. 387–405.
Scholtens, B. P. C., *Bosnegers en overhead in Suriname. De ontwikkeling van een politieke verhouding 1651–1992* (Paramaribo: Afdeling Cultuurstudies, 1994).
Scott, J. C., *Domination and the Art of Resistance: Hidden Transcripts* (New Haven, CT: Yale University Press, 1990).

Scott, J. C., *Seeing Like a State: How Certain Schemes to Improve The Human Condition Have Failed* (New Haven, CT: Yale University Press, 1998).

Seng, L. K., *Making and Unmaking the Asylum: Leprosy and Modernity in Singapore and Malaysia* (Petaling jaya: Strategic Information and Research Development Centre, 2009).

Sheridan, R. B., *Doctors and Slaves: A Medical and Demographic History of Slavery in the British West Indies, 1680–1834* (Cambridge: Cambridge University Press, 1985).

Smith Kipp, R., 'The evangelical uses of leprosy', *Social Science and Medicine* 39 (1994), pp. 165–78.

Snelders, S., '"Kapers van kennis". De rol van een boekaniersgeleerde in de circulatie van kennis over ziekten en geneesmiddelen in de tropen', *Studium* 2 (2009), pp. 55–64.

Snelders, S., *Vrijbuiters van de heelkunde. Op zoek naar medische kennis in de tropen 1600–1800* (Amsterdam: Atlas, 2012).

Snelders, S., 'Leprosy and slavery in Suriname: Godfried Schilling and the framing of a racial pathology in the eighteenth century', *Social History of Medicine* 26 (2013), pp. 432–50.

Snelders, S., and F. J. Meijman, *De mondige patient. Historische kijk op een mythe* (Amsterdam: Bert Bakker, 2009).

Spapens, P., *Gwasi siki. Levensverhalen van Surinaamse mensen die lepra hebben gehad* (Tilburg: Pix4Profs, 2012).

Stipriaan Luïscius, A. A. van, 'Surinaams contrast. Roofbouw en overleven in een Caraïbische plantage-economie' (PhD thesis, Vrije University Amsterdam, 1991).

Stoler, A. L., *Carnal Knowledge and Imperial Power: Race and the Intimate in Colonial Rule* (Berkeley, CA: University of California Press, 2002).

Superlan, P., 'The Javanese in Surinam: Ethnicity in an Ethnically Plural Society' (PhD thesis, University of Illinois, 1978).

Themen-Sliggers, M., 'De maatschappelijke aspecten van lepra', *OSO* 22 (2003), pp. 104–11.

Thoden van Velzen, H. U. E., and W. Hoogbergen, *Een zwarte vrijstaat in Suriname. De Okaanse samenleving in de achttiende eeuw* (Leiden: KITLV Uitgeverij, 2011).

Thomas, K., *Man and the Natural World: Changing Attitudes in England 1500–1800* (London: Allen Lane, 1983).

Trouillot, M.-R., *Silencing the Past: Power and the Production of History* (Boston, MA: Beacon Press, 1995).

Vaughan, M., *Curing Their Ills: Colonial Power and African Illness* (Cambridge: Polity Press, 1991).

Veldhuyzen, W. F. H., *Honderd en vijftig jaar pokken preventie* (Amsterdam: Scheltema and Holkema, 1957).
Vernooij, J., *'Barmhartigheid een levensprogram. Zusters van liefde van Tilburg 100 jaar in Suriname'* (Paramaribo, n.p., 1994).
Vernooij, J., *De rooms-katholieke gemeente in Suriname. Handboek van de geschiedenis van de Rooms-Katholieke Kerk in Suriname* (Paramaribo: Leo Victor, 1998).
Vernooij, J., 'Een opvallende relatie. De rooms-katholieke kerk en lepra in Suriname', *OSO* 22 (2003), pp. 62–8.
Vijftig jaren Hofbauer liefdewerk 1890–1940 (Rotterdam: Secretariaat der Surinaamsche Missie, 1940).
Vollset, M., 'Globalizing Leprosy: A Transnational History of Production and Circulation of Medical Knowledge 1850–1930' (PhD thesis, University of Oslo, 2013).
Waal Malefijt, A. H. de, 'The Javanese population of Surinam' (PhD thesis, Colombia University, 1964).
Ward, J. R., *British West Indian Slavery, 1750–1834* (Oxford: Clarendon Press, 1988).
Weaver, K. K., *Medical Revolutionaries: The Enslaved Healers of Eighteenth-Century Saint Domingue* (Urbana, IL: University of Illinois Press, 2006).
Wekker, G., *The Politics of Passion: Women's Sexual Culture in the Afro-Surinamese Diaspora* (New York: Columbia University Press, 2006).
Wooding, C. J., 'Winti: Een Afroamerikaanse godsdienst in Suriname. Een cultureel-historische analyse van de religieuze verschijnselen in de Para' (PhD thesis, University of Amsterdam, 1972).
Worboys, M., 'The emergence of tropical medicine: A study in the establishment of a scientific specialty', in G. Lemaine, R. Macleod, M. Mulkay, and P. Weingart (eds.), *Perspectives on the Emergence of Scientific Disciplines* (The Hague: Mouton, 1976), pp. 75–98.
Worboys, M., 'Tropical diseases', in W. F. Bynum and R. Porter (eds.), *Companion Encyclopedia of the History of Medicine*, vol. 2 (London: Routledge, 1993), pp. 512–36.
Worboys, M., 'The colonial world as mission and mandate: Leprosy and empire, 1900–1940', *Osiris* 15 (2000), pp. 207–18.

Secondary sources, unpublished manuscripts

Bergen, L. van, 'De vreeselijkste van alle kwalen Lepra in Nederlands-Indië 1815–1942' (Royal Netherlands Institute of Southeast Asian and Caribbean Studies, Leiden, 2015).

Menke, H. E., 'The Landré Family: Drama and Scientific Concept in a Slave Colony Ravaged by Leprosy' (Rotterdam, 2010).
Reyme, M., 'Give Ex-Leprosy Patients a Voice' (unpublished transcripts interviews, Anton de Kom Universiteit van Suriname, Paramaribo, 2013–2014).
Vernooij, J., 'De rooms-katholieke missie in Suriname ten tijde van Mgr. Grooff (1826–1853)' (Nijmegen, 1976).
Worboys, M., '"An Imperial Danger": Leprosy and Contagion, 1860–1900' (University of Manchester, 2004).

Index

adventurer-scientists 31
aetiology 32–3
  see also climate and leprosy; diet; sexuality; transmission
Afro-Surinamese
  after Emancipation 120–1
  terminology 16n30, 38n9
  see also slavery
Afro-Surinamese medicine
  belief systems 52, 78, 84–6, 98, 122–3, 190, 199–211, 249
  dresiman (black surgeon) 28, 47–8
  female healers 36, 48, 80–2
  healing methods 36, 78–84, 89n9
  knowledge 35, 78–9, 82–3, 127
  lukumen 81–2, 87, 208–9
  obiah men 201, 208–9
  wisimen 201, 208–9
  see also treef; Winti
Amerindians 23, 81–2, 147, 149
Andel, T. van 80
Anderson, W. 9, 221
Andreia 107
Arnold, D. 6

Asch van Wijk, T. van 150–2
authoritarian modernism 7, 119, 162

Bakker, R. P. 147, 155
Bankole, K. 79
Barrère, P. 6
Batavia leprosy asylum 7, 50, 78, 93–111, 122, 136, 145–50, 156, 219, 224, 226–7
  and compulsory segregation polices 56–71
  population figures 102–5, 124
Beek, J. P. ter 53, 64
Benjamins, H. 203, 205
Bethesda 7, 146, 155–6, 191, 219, 223–6, 230–4
  population figures 164–6
  treatment 236–9
Bijker, K. 97
Blom, Annemarieke 10
Blom, Anthony 35, 81
boasie 22, 37–8n3, 83
Boeck, C. W. 68, 130–1, 134
Bosman, W. 85

British Caribbean 132, 143–5, 152, 220–1
British Indians 121, 123, 149, 161, 167, 169, 173, 212–13, 223–4, 226, 248
  belief systems 199, 211
  in Bethesda 234
  leprosy among 122, 132, 163, 166, 200, 211
British occupation of Suriname 44–5
Broers van Dort, T. 153
Bronbeek 124–9, 131–2
Büchner, W. 53
Buckingham, J. 8, 10, 212, 223
Bueno de Mesquita, S. J. 230, 238

Calkar, H. van 108–9
Cantz'laar, P. 101
Carsten, B. 132–3
Catholics in Suriname 8, 95–7, 102, 142–4, 146, 149–50, 219–20, 225–30
  in Batavia 93–4, 96–102, 105–11, 147–50, 156
  see also Majella leprosy asylum
CGOT (Committee for Medical Investigation and Supervision) see medical care in Suriname
chaulmoogra therapy 176, 235–8
children see leprosy sufferers
Chinese 121
  leprosy among 122, 132, 163
'Christian leper identity' 94, 98–9, 228, 233–4
climate and leprosy 32–3, 53, 64–5, 69, 125, 134–5, 223
Colijn, H. 187–8
Coll, C. van 82

Collegium Medicum see medical care in Suriname
Colonial Estates 122, 136, 146, 150–1, 163, 167–70, 178–80, 186, 188–9, 191, 204, 228–30
colonization of the body 3, 6, 162, 248
Committee of Investigation 57–62, 65, 67, 124, 135–6
compulsory segregation 3, 7, 28, 58–63, 119, 135–6, 142, 145, 162–3, 167–93, 205, 248
  see also edicts in Suriname: 1790, 1830, 1929; 'Great Confinement'
contagiousness 22, 26, 34, 49, 67–9, 120, 125, 129–36, 142, 144, 152–3, 175, 207–9
  see also heredity; sexuality; transmission
contestation 9, 185–6, 189–90, 193
  infrapolitics 9, 163, 190, 250
  leprosy asylums 94–5, 106, 108–11, 147–9, 186, 226–7, 233–4
  Suriname 179, 248
  treatment 237–40
Cool, P. 189, 226, 229–30, 233, 237
Coster, A. 85
Crosby, A. 5
cures see Afro-Surinamese medicines: healing methods; treatments in colonial medicine

Dahlberg, C. G. 83
Damien (J. de Veuster) 8, 105, 142, 144, 147, 155
Danielssen, D. C. 68, 130–1, 134
Davis, N. Z. 38n7

DDS (diaminodiphenyl sulphate)
    see sulfone therapy
Deutschbein, L. L. A. 68, 107
diagnosis 28, 34–5, 126–8
diet 32–3, 52–3, 64–5, 69, 207–9,
    212, 218n43
Donders, P. 98, 105, 109–10,
    146–8, 155
Douglas, M. 1
Drognat Landré, C. L. 58, 61, 67,
    129–36, 152
Dutch East Indies 128–9, 133, 152,
    168–9, 174, 177, 210, 213–14
    edict of 1865 132

edicts in Suriname
    1728 25
    1761 25–7
    1764 26–7
    1780 28, 33
    1790 28–9, 33
    1791 29, 33
    1824 57
    1830 43, 46, 57–8, 146,
        166–7, 169
    1831 61
    1845 63
    1853 67
    1855 67
    1929 161, 172, 178, 186–8, 205
Edmond, R. 1, 9, 221
elephantiasis 43, 45, 49, 57,
    62, 66, 70
    see also filariasis
Emma, Queen-Mother 151, 231
epidemiological transition in the
    Caribbean 5
Essed, W. 205

Essequibo 24–5
Esther Foundation 192

Fermin, P. 26, 82–3, 127
Fermor, P. L. 220–1
filariasis 167, 230
    see also elephantiasis
Flu, P. C. 174–5, 232
Foucault, M. 7, 9, 43, 177, 222
Friedman, S. 69, 107
Fuchs, C. 53

Gallandat, D. H. 41n42
gender 185–6, 225, 227, 231–2
Gravenhorst, C. N. G. 56
'Great Confinement' 7, 43, 66,
    69–71, 88
Grooff, J. 51, 99, 101–2, 105, 107–8
Groot, H. 147
Groot-Chatillon 7, 146, 151–2,
    155–6, 170, 186, 192, 212–13,
    219, 223–7, 229–30, 232, 234
    population figures 164–6, 211, 224
    treatment 235–9
Gussow, Z. 3–4

Hansen, A. 136, 173
Hardt, M. 247
Harrison, M. 5
Hasselaar, A. van 45, 49–55, 61–2,
    88, 131
    visit to Batavia 50, 99–100
Heekeren, E. L. Baron van 102
Heinink, G. J. 107–8
Hek, J. H. 97
heredity 64, 68–9, 120, 130–4
Hering, C. 45, 54–6, 62, 83, 101
Herrnhutters see Moravians

Index

Herskovits, F. 87
Herskovits, M. 87
Heshuyzen, F. van 48–9
Hille, J. 62
Hindus *see* British Indians
Hoefte, R. 9, 179
homeopathy *see* Hering, C.
Horst, L. van der 24, 96
humoral medicine 32–3, 35

Idenburg, A. 168–9
Idsinga, W. van 135
imperial danger 3–4
Imperial Tropical Medicine 173, 200
indentured labour 2, 119–21
infrapolitics *see* contestation
Ingliss, K. 8, 10
International Leprosy Conference
   Bergen 1909 175, 177
   Berlin 1897 144, 152, 177
   Strasbourg 1923 175–7

Javanese 121, 123, 173, 213,
   223–4, 248
   belief systems 199, 211, 213–14
   leprosy among 200, 213
Jewish savants 81
Jews 46, 53, 85–6, 133–4, 249
Juliana, Princess 231

Kanter, P. de 65–6
Keil, E. G. 186, 230, 232
Kielstra, J. 179, 232
Kiple, K. 5
Kleinman, A. 97
kokobe 83, 87
Kristeva, J. 1
Kuhn, F. 45, 47, 49–50, 87
Kwatta 55–6

Lammens, A. 81
Lampe, P. H. J. 86, 171–4, 177–8,
   190, 201, 205–8, 210, 224, 226,
   229, 236–7
Landré, C. 58, 61–2, 65, 67, 87,
   126–7, 129–31, 134, 137
late colonial capitalism 3, 119,
   179, 191
Leent, F. J. van 133–4
Lely, C. 166–7, 170, 203
Lemmens, F. 228, 235
Lens, P. 148–9
leprosy asylums 7–10, 46, 135, 142,
   145, 150, 152–6, 167–70,
   178, 219–40
   complex macrocosms 223
   disciplinary power 10, 177,
   219–22
   population figures 102–5, 124,
   163–5, 181–4, 227, 231–2, 234
   see also Batavia leprosy asylum;
   Bethesda; Groot-Chatillon;
   Majella leprosy asylum;
   Voorzorg
leprosy sufferers 12–13n1, 148–9
   agency 8
   children 57, 63–4, 148, 168, 180–
   1, 225, 234
   estimated numbers 65, 70–1, 153,
   169–71, 184–5, 206
   European 29, 51–2, 82, 200
   marriage 52, 168, 220, 225
   marronage 226, 229–30, 240
   stigmatization 1–2, 10–11, 35, 52,
   161, 169, 189, 193, 248–50
   see also contestation; leprosy
   asylums
Lier, R. van 120, 122

Ludwig, J. F. 31
Lutherans *see* Protestants

Magnée, C. J. 108–9, 147
Majella leprosy asylum 7, 146,
    154–6, 168, 191, 206, 212, 219,
    223–4, 226–32
  population figures 164–6
  treatment 236–7
Manson, P. 173
Maroons 23–4, 66, 81
marriage *see* leprosy sufferers
May, T. 86, 175, 179, 205
MCC (Middelburgsche
    Commercie Compagnie)
    31, 41n42
medical care in Suriname 26, 46–8,
    123–4, 172
  CGOT (Committee for Medical
    Investigation and Supervision)
    63–5, 67, 123
  Collegium Medicum 25, 31, 47,
    49, 57, 63, 99
  Medical Committee 123
  surgeons 26, 47–8, 124
Military Hospital of Paramaribo
    31, 170, 172, 188, 201,
    235, 237
missionary societies 4, 8, 144–5,
    168, 176, 219–22, 249
  see also Catholics in Suriname;
    Moravians; Redemptorists;
    Sisters of Love
modern colonial policies 7, 119,
    162, 214–15
  see also authoritarian modernism;
    late colonial capitalism;
    modern leprosy policies

modern leprosy policies 7–8,
    162, 179, 192–3, 199, 202
Moravians 55, 94, 101–2,
    108–10, 143, 149, 155,
    231, 233
Muir, E. 176, 236

Native Americans *see* Amerindians
Negri, T. 247
Netherlands, The
  medical discussions 53, 64–6, 69,
    125, 129–37
  return of leprosy 34, 120, 124–6,
    128, 247
  state leprosy asylum *see*
    Veenhuizen
  see also parliament of the
    Netherlands
New-Bethesda 232–4
  see also Bethesda

Ooijkaas, 69, 107–8
outpatient clinic 181
outpatient school 181, 188, 191

Panhuys, W. van 45–6
parliament of the Netherlands 136,
    170–1, 173, 178, 180, 186–7,
    191, 201–2, 226, 228
Penard, A. 83, 87
Penard, F. 83, 87
Peters, O. A. 143–6, 149–50,
    153, 171
plantations 2, 23, 65, 81, 120–1,
    179, 247–8
  management of leprosy sufferers
    26, 50, 60–1
  medical care 47–8

Index

police 57, 63–4, 67, 171, 175, 180, 202, 226
population of Suriname 23–4
Protestants 34, 142–3, 151–2, 155, 219–20, 231–4
  see also Bethesda; Moravians
Pruys van der Hoeven, C. 64–6
psoriasis 55, 62, 87, 127, 207

Quassie 82
Quinian, S. 6

racial pathology 23–4, 33–7, 48, 50, 53, 65, 163, 247–9
racism
  colonial medicine 5–6, 25, 45, 49, 55, 173, 173, 176–7, 247–9
  leprosy 33–7, 50–3, 133–4, 145–6, 165–6, 174, 200, 233, 237–8
  see also racial pathology
Redemptorists 146–7
Rijk, J. C. 64, 102
Robertson, J. 4, 10, 222
Rodschied, E. K. 25
Rogers L. 176–8, 210
Rolander, D. 21–2, 27, 31
Royal College of Physicians (British) 68, 131–2, 135, 144

Salomons, A. 123, 153–4, 163, 166, 172
Samweri 201
Sanders, C. H. 135
Savornin Lohman, M. Jonkheer de 143–4, 150
Schaick, C. van 51
Schepers, J. G. 108
Schiebinger, L. 5

Schilling, G. W. 22–3, 30–7, 38n7, 48, 50, 52–3, 64–5, 69, 79–82, 84, 130–1, 134, 247
Schimmelpenninck van der Ooije, J. Baron 151–3
Schlörholtz, E. P. 62
Schönfeld, K. 69
Schuitemaker, F. P. 172, 175, 178, 186, 229, 236
Scott, J. 9, 147, 162, 189–90
sexuality 33–4, 48, 50–3, 63, 65, 109–10, 147, 149, 166, 200, 225, 234
  interracial 33–4, 42n56, 49, 51–2, 55–6, 130, 133, 175, 248
Sheridan, R. 5
Silla, E. 8
Simons, C. 206
Simons, R. D. P. G. 37–8n3, 85–6, 202, 211
Sisters of Love 154–5, 228
slave owners 6, 27, 29–30, 61, 64, 70, 78, 86, 124, 179
slavery 2, 23, 27, 57, 70, 119–20, 124, 247
  see also slaves
slaves
  medical examinations 31
  mortality rates 27
  prices 27–8
  trade 31
Smith-Kipp, R. 9, 221
Society of Suriname 23, 29, 44
sources
  bottom-up perspective 9–10, 80, 88, 249
  silence 9, 79–80
Stedman, J. G. 24, 26, 84–5

stigmatization *see* leprosy sufferers
Storm van 's-Gravesande, L. 24–5
sulfone therapy 191–2, 238–9
surgeons *see* Afro-Surinamese medicine; medical care in Suriname
symptoms of leprosy 2, 34–5, 50, 87–8, 127

Teenstra, M. D. 88
terminology of leprosy 22, 32, 38n4, 45, 68, 83, 127, 214
Trackanen, N. 125–6, 128
transmission 32, 173, 200–1, 205–7, 209–11, 229–30, 247–50
  see also contagiousness; heredity; sexuality
treatments in colonial medicine 35–6, 53–6, 176, 235–9
  see also Afro-Surinamese medicine; chaulmoogra therapy; sulfone therapy
treef 78, 84–8, 146, 199–209, 214

Uhlig C. 108, 134

Vaughan, M. 98
Veenhuizen 129, 131, 135
Veer, A. de 95–6, 99
Vinkhuizen, H. J. 134–5
Virchow, R. 53
Voorzorg 7, 28–9, 56, 78, 94–7
Vries-Bruins, A. de 187–8, 226
Vrolik, G. 64–6

Weijden, M. van der 97–9
Wekker, G. 84
Wennekers, P. A. 95–6
WIC (West India Company) 23
Wijnkoop, D. 187
Wilhelmina, Queen 191
Willemsen, J. 28–9, 96–7
Winti 81, 84, 201
Woensel, P. van 28–9, 33–5
Wolbers, J. 105
Wolff, J. W. 174, 215n4
Worboys, M. 8
Wright, H. P. 144
Wulfingh, W. 144, 146, 149–52, 154, 156

yaws 25, 30–1, 54, 213

Zaalberg, H. H. 151, 155, 231

EU authorised representative for GPSR:
Easy Access System Europe, Mustamäe tee 50,
10621 Tallinn, Estonia
gpsr.requests@easproject.com

www.ingramcontent.com/pod-product-compliance
Ingram Content Group UK Ltd.
Pitfield, Milton Keynes, MK11 3LW, UK
UKHW021903240326
469329UK00005B/10